Collateral Damage

Echoes of War: The Russia-Ukraine Conflict and Its Global Ripples, Volume 5

SANDEEP CHAVAN

Published by SANDEEP CHAVAN, 2024.

COLLATERAL DAMAGE

First edition. December 19, 2024.

ISBN: 979-8230960065

Written by SANDEEP CHAVAN.

Table of Contents

This book is dedicated to the countless victims of war whose lives have been irreparably altered by conflict, and to the resilient individuals working tirelessly to heal damaged environments, rebuild communities, and provide hope to those affected.

It is also dedicated to the ethical hackers and cybersecurity experts who stand on the digital frontlines, safeguarding critical systems and protecting the integrity of information in an increasingly volatile cyber landscape.

Your courage, perseverance, and commitment to recovery and security inspire us all. May this book serve as a testament to your efforts and a call to action for a more resilient and compassionate world.

"In the ashes of destruction lie the seeds of resilience, and in the shadows of conflict, the light of innovation can guide us toward hope and recovery."— Anonymous

Disclaimer

This book, *Collateral Damage: The Russia-Ukraine War's Toll on Earth and Cyberspace*, is intended for informational and educational purposes. The views and opinions expressed herein are those of the author and contributors and do not necessarily reflect the official policies or positions of any government, organization, or institution.

While every effort has been made to ensure the accuracy and completeness of the information presented, the author and publisher make no guarantees or warranties regarding the reliability, applicability, or completeness of the content. Readers are encouraged to independently verify facts and consult additional sources for a comprehensive understanding of the topics discussed.

The book addresses complex and evolving subjects related to environmental impacts, cybersecurity, and modern warfare. As such, some examples, case studies, or scenarios are presented for illustrative purposes and may not reflect the full context or outcomes.

The mention of specific organizations, companies, or individuals is not intended to imply endorsement, affiliation, or criticism. Similarly, the inclusion of technologies or strategies does not constitute professional or legal advice.

The author and publisher disclaim any liability for errors, omissions, or any adverse consequences arising from the use or interpretation of the information contained in this book. Readers are encouraged to approach these topics with critical thinking and seek expert guidance where necessary.

This book is a work of research and analysis, and its purpose is to foster awareness, discussion, and action on the interconnected challenges of war, technology, and the environment. It is not intended to promote or advocate any political or ideological position.

Thank you for engaging with this work responsibly and thoughtfully.

Foreword

The Russia-Ukraine war stands as a stark reminder of the multifaceted challenges modern conflicts pose to our world. Beyond the immediate human toll, this war has revealed how battles extend into two critical and interconnected realms: the environment and cyberspace. The consequences of these dual dimensions of warfare ripple far beyond the conflict zone, affecting ecosystems, digital infrastructure, and global security.

This book, *Collateral Damage: The Russia-Ukraine War's Toll on Earth and Cyberspace*, delves into the often-overlooked impacts of war, focusing on environmental devastation and the vulnerabilities exposed in the digital age. With meticulous analysis and real-world examples, it explores the challenges of recovery and the opportunities to rebuild with resilience and sustainability at the forefront.

The author brings a unique and holistic perspective to these pressing issues, weaving together insights from environmental science, cybersecurity, and international cooperation. The result is a work that not only documents the scars left by war but also charts a path forward—one grounded in innovation, collaboration, and hope.

This book is more than an exploration of the conflict; it is a call to action for governments, organizations, and individuals. It underscores the urgency of addressing the interconnected challenges of environmental and cyber resilience and inspires readers to think critically about the role we all play in shaping a secure and sustainable future.

As you read through the pages of *Collateral Damage: The Russia-Ukraine War's Toll on Earth and Cyberspace*, I encourage you to reflect on the lessons it offers. The challenges may be daunting, but they also present an opportunity to innovate and collaborate for a better tomorrow. This is not just a story of loss; it is a blueprint for recovery, resilience, and renewal.

Pradeep Chavan
Chemical Engineer

Preface

War has always been a destructive force, but the modern battlefield extends far beyond the physical borders of nations. The Russia-Ukraine conflict has laid bare the devastating consequences of warfare on both the natural environment and the digital world, revealing vulnerabilities and interconnections that demand our urgent attention. This book, *Collateral Damage: The Russia-Ukraine War's Toll on Earth and Cyberspace*, explores the often-overlooked dimensions of war—its impact on ecosystems and the cybersphere—and the lessons we must learn to safeguard our planet and our future.

The inspiration for this book arose from the need to understand the broader implications of modern conflicts. While much attention is paid to human casualties and economic losses, the scars left on our environment and digital infrastructure often remain hidden. These are not isolated issues but interconnected challenges that influence global stability, resource security, and the way we perceive and engage with technology and nature.

The chapters in this book delve into the dual toll of war, from the environmental devastation wrought by bombings, deforestation, and chemical spills to the cyberattacks targeting critical infrastructure and manipulating global narratives. It examines the resilience of ecosystems and digital systems, the ethical dilemmas of leveraging technology in warfare, and the collaborative efforts needed to rebuild and restore after conflict. Through case studies, expert insights, and a forward-looking perspective, the book seeks to shed light on these pressing issues and spark meaningful discussions on prevention and recovery.

As an author, my aim is to present a balanced and comprehensive exploration of these topics, drawing connections between the physical and digital worlds and emphasizing the importance of sustainable and ethical solutions. This book is not just about documenting the past but about charting a course toward a more resilient and interconnected future.

I would like to express my gratitude to everyone who contributed their knowledge, experiences, and perspectives to this project. This work is as much a reflection of their efforts as it is of my own. To the readers, I hope this book inspires you to think critically about the challenges we face and the role we can all play in addressing them.

The Russia-Ukraine conflict is a stark reminder that the impacts of war transcend borders, leaving a lasting legacy on the Earth and the systems we depend on. By understanding these impacts, we can take the first steps toward preventing them and building a world that prioritizes resilience, collaboration, and sustainability.

Sincerely,

Sandeep Chavan

Introduction

Setting the Stage: The Dual Toll on Earth and Cyberspace

War has always been a harbinger of destruction, but in today's interconnected world, its consequences extend far beyond the boundaries of traditional battlefields. The Russia-Ukraine war has revealed the multifaceted nature of modern conflicts, where destruction is no longer confined to physical landscapes. It is a war that reverberates through both the tangible world of ecosystems and the intangible realm of cyberspace. This dual toll—on Earth and in cyberspace—has emerged as a defining characteristic of contemporary warfare, highlighting the expanded scope of human suffering and instability.

In the physical realm, war scars the land and poisons the air and water. Bombs crater the earth, forests are scorched, and rivers are polluted with oil and chemicals. The environmental devastation wrought by traditional warfare is immediate and painfully visible, leaving behind a legacy of barren landscapes and disrupted ecosystems. Yet, in today's digital age, the battlefield has expanded to include cyberspace—a realm where attacks are stealthy, pervasive, and often invisible to the naked eye. Cyberattacks infiltrate networks, disrupt critical infrastructure, and erode the trust that underpins societies. While a forest fire caused by a bombing raid is tangible and undeniable, the crippling of a nation's power grid by malware is just as impactful, though far less visible.

These two domains—natural ecosystems and digital infrastructures—are often viewed as separate entities, but the Russia-Ukraine conflict has shown that they are deeply interconnected in their consequences. The environmental degradation caused by attacks on oil fields, power plants, and industrial facilities finds its echo in the cyber realm, where sabotaged systems disrupt food supplies, healthcare, and communication networks. The vulnerabilities exposed in one domain often amplify the crises in the other, creating a ripple effect that extends far beyond the borders of the warring nations.

The Russia-Ukraine war exemplifies how modern warfare amplifies human suffering through these dual dimensions. Burning oil fields and devastated agricultural land mirror the compromised digital systems that wreak havoc on economies and societies. Sabotaged power grids not only leave millions in darkness but also exacerbate environmental crises by forcing reliance on backup systems that are inefficient and polluting. Weaponized misinformation spreads fear and division, destabilizing communities much like the destruction of forests destabilizes ecosystems.

This convergence of environmental and cyber impacts underscores the necessity of addressing the toll of war holistically. Each dimension of destruction shapes the lives of millions in ways that are often underreported or overlooked. Recognizing the interconnectedness of these impacts is crucial for understanding the true cost of modern conflicts and for forging a path toward recovery and resilience in a world that increasingly blurs the line between the physical and the digital.

The Russia-Ukraine war is not just a test of military might; it is a stark warning about the evolving nature of warfare and its capacity to leave behind a legacy of destruction that transcends borders, generations, and domains.

The Russia-Ukraine War as a Modern Conflict: Combining Traditional Warfare with Cyber and Environmental Repercussions

The Russia-Ukraine war has emerged as a harbinger of the complexities of modern warfare, a proving ground for new strategies and tools that extend the battlefield beyond traditional combat zones. It is a conflict that fuses the brutality of conventional military operations with the precision and reach of cyberattacks, creating a multi-dimensional theater of war. Missiles and tanks operate alongside malware and disinformation campaigns, forming a layered battleground where the metrics of victory and defeat are no longer limited to captured territory but include destabilized networks and devastated ecosystems.

Traditional warfare in this conflict has left behind an extensive trail of environmental destruction. The bombing of industrial zones has triggered chemical spills that contaminate rivers, lakes, and soil, jeopardizing both biodiversity and human health. Forests, some of which stood as natural sanctuaries for centuries, have been reduced to ash—not for their resources, but as unintended casualties of relentless military operations. Fields of crops have been set ablaze or rendered unusable due to landmines and shell fragments, creating food insecurity and disrupting local and global agricultural supply chains. These acts of environmental devastation are not only immediate in their consequences but also carry long-term repercussions that will haunt the region for generations.

Parallel to the visible carnage is the less tangible but equally disruptive dimension of cyber warfare. Unlike the physical scars left by bombs and missiles, cyberattacks strike invisibly but effectively, targeting critical infrastructure such as power grids, banking systems, and communication networks. In Ukraine, blackouts caused by cyberattacks have plunged millions into darkness, compromising healthcare systems, halting industries, and leaving people vulnerable amid the chaos. Economic systems have faced instability as hackers infiltrate banking networks and manipulate financial data. Beyond infrastructure, the weaponization of misinformation and propaganda has eroded trust within communities, destabilizing societies and sowing division far beyond the battlefield.

What makes this conflict particularly unsettling is its ability to ripple outward, touching lives and systems far removed from the epicenter of combat. In an increasingly interconnected world, cyberattacks launched in Eastern Europe can paralyze systems in countries across the globe, disrupting supply chains, interfering with governance, and spreading disinformation at a scale previously unimaginable. Similarly, the environmental devastation in Ukraine does not remain confined to its borders. Pollutants leaching into waterways can traverse international boundaries, and the destruction of forests contributes to global challenges such as climate change by reducing carbon sequestration and increasing greenhouse gas emissions.

This war serves as a stark reminder that the impacts of modern conflict are neither linear nor localized. It underscores the evolving nature of warfare, where the theater of combat extends into domains that were once deemed peripheral—nature and cyberspace. The interdependence of these domains has created a conflict that is as much about survival in the digital realm as it is about controlling physical territory. The Russia-Ukraine war exemplifies how modern conflicts do not just challenge military strategies but also strain the resilience of ecosystems and digital infrastructures, leaving humanity to grapple with a legacy of destruction that transcends borders and generations.

Purpose of the Book: Understanding the Unseen Impacts of War on Ecosystems and Digital Domains

War is often framed in terms of immediate and tangible destruction: the lives lost, cities reduced to rubble, economies in disarray, and territories seized or defended. Yet beneath these visible scars lie the quieter but no less devastating impacts on the natural world and the digital frameworks that underpin modern life. This book seeks to illuminate these hidden dimensions of conflict, focusing on the environmental devastation and cyber repercussions that are frequently overshadowed by the more dramatic, headline-grabbing aspects of war.

The Russia-Ukraine war offers a stark example of how modern conflicts have evolved, extending their reach far beyond the battlefield. The environmental toll of the war serves as a sobering reminder of the fragility of the natural systems upon which all life depends. Forests, rivers, and agricultural lands have become unintended casualties, bearing the brunt of bombings, industrial explosions, and military activity. These ecological disruptions are not merely local tragedies; they have global implications, from exacerbating climate change to threatening biodiversity and food security.

At the same time, the rise of cyber warfare represents a fundamental shift in the way nations and non-state actors engage in conflict. Unlike physical battles, cyberattacks often unfold invisibly, targeting critical infrastructure, eroding public trust, and destabilizing economies. The Russia-Ukraine war has demonstrated how digital tools can be weaponized to paralyze nations and disrupt global systems, exposing vulnerabilities in our interconnected world. Whether it is a blackout caused by malware or the manipulation of information to sow discord, the consequences of cyber warfare are far-reaching and insidious.

Through this exploration, the book aims to provide a comprehensive understanding of these less-visible dimensions of the Russia-Ukraine war. By delving into the environmental and cyber impacts, it seeks to uncover the broader implications for global security, environmental sustainability, and the resilience of our societies. Case studies and expert insights will illustrate how these dual impacts—on Earth's ecosystems and cyberspace—reflect the evolution of warfare and the unprecedented challenges humanity must address in mitigating their consequences.

In highlighting these issues, this book hopes to inspire meaningful dialogue about accountability, recovery, and prevention. It seeks to challenge the traditional narratives of war that focus solely on immediate human and economic costs, urging readers to consider the long-term collateral damage inflicted on the natural world and digital systems.

Ultimately, this book argues that war is no longer fought solely with guns and tanks but also with the ecosystems that sustain us and the digital networks that connect us. Understanding this new reality is critical, not only for addressing the consequences of current conflicts but also for preventing similar devastation in the future. As we examine the dual toll of the Russia-Ukraine war, the hope is to foster greater awareness, inspire actionable solutions, and advocate for a more holistic approach to conflict resolution that prioritizes the protection of both our planet and our digital future.

Part I: The Environmental Fallout

Chapter 1: The Battlefield and Nature

War has always left scars on the land it touches, but the Russia-Ukraine conflict has brought into sharp focus the devastating toll that modern warfare exacts on nature. Beyond the human tragedies and urban destruction lies a quieter, often overlooked casualty: the environment. Forests burn, rivers are poisoned, and the air becomes thick with pollutants from relentless bombings, shelling, and industrial ruin. These environmental damages are not merely collateral; they are enduring reminders of how war disrupts ecosystems, damages biodiversity, and alters landscapes for decades.

The battleground in Ukraine is not just a clash of armies and ideologies—it is a direct assault on the natural world. From toxic chemicals leaking into rivers to the destruction of fertile lands and the disruption of delicate ecosystems, the environmental repercussions are profound and far-reaching. This chapter explores the immediate impacts of the war on forests, rivers, and air, shedding light on the silent but significant devastation unfolding in the natural world. Through specific case studies and analysis, it delves into the ecological price of war, a cost that will continue to ripple long after the guns fall silent.

1.1 The Immediate Environmental Cost of War

War has always exacted a steep toll on the environment, but modern conflicts amplify this destruction with devastating precision and scale. The environmental consequences of warfare are often overshadowed by the immediate focus on human casualties, economic losses, and geopolitical shifts. Yet, the natural world becomes an unintended victim of human aggression, bearing scars that last decades—if not centuries. From the destruction of forests and rivers to the pollution of air and soil, the environmental toll of war is both immediate and far-reaching, often creating long-term crises that transcend borders.

The Direct Impact on Forests

Forests are frequently caught in the crossfire of modern warfare. During the Russia-Ukraine war, vast tracts of woodlands have been reduced to ash due to bombings, missile strikes, and military encampments. Trees, which take decades to grow, are destroyed in moments, leading to a loss of biodiversity, carbon sequestration, and critical habitats for wildlife. In some cases, forests are deliberately targeted for tactical reasons—clearing them to deny cover to the enemy—or unintentionally destroyed as collateral damage.

Forests not only act as carbon sinks, mitigating climate change, but also provide resources and livelihoods for local communities. When they are damaged or destroyed, the effects ripple through the ecosystem, impacting everything from soil health to water retention. In Ukraine, regions rich in forest cover, such as the Carpathians and parts of the Donbas, have seen significant deforestation, disrupting delicate ecological balances and leaving barren landscapes in place of once-thriving ecosystems.

Rivers and Waterways under Threat

Rivers, often lifelines for agriculture, industry, and human consumption, are highly vulnerable during conflicts. The bombing of industrial facilities near rivers releases hazardous chemicals, heavy metals, and oil into the water, contaminating ecosystems and threatening both human and aquatic life. In Ukraine, key waterways like the Dnipro River have faced significant environmental stress due to nearby military activity. Chemical spills, wastewater leaks, and destroyed infrastructure have resulted in polluted rivers, rendering them unsafe for drinking, irrigation, or sustaining aquatic ecosystems.

Water pollution during wartime has a cascading effect. Contaminated rivers flow downstream, spreading pollutants far beyond the conflict zones. This not only affects local communities but also poses risks to neighboring regions and countries that rely on shared water resources. Furthermore, the disruption of water supplies due to damaged infrastructure leaves civilians vulnerable to disease outbreaks and water scarcity, compounding the humanitarian crisis.

The Air We Breathe: Pollution from Conflict

Modern warfare releases an alarming amount of pollutants into the air, posing immediate and long-term health risks. Explosions from bombs, shelling, and missile strikes fill the air with fine particulate matter, heavy metals, and toxic chemicals, creating a hazardous environment for both combatants and civilians. Burning oil depots, industrial facilities, and military equipment contribute to massive releases of carbon dioxide, sulfur dioxide, and other greenhouse gases, accelerating climate change.

During the Russia-Ukraine war, cities under heavy bombardment, such as Mariupol and Kharkiv, have experienced significant air pollution levels, making it unsafe for residents to breathe. The smoke and debris from destroyed buildings and infrastructure linger in the atmosphere, impacting air quality over vast areas. Prolonged exposure to such conditions can lead to respiratory illnesses, cardiovascular issues, and long-term health problems for those living in or near conflict zones.

Modern Military Operations: Accelerating Environmental Destruction

The scale and intensity of modern military operations exacerbate environmental destruction far beyond what was seen in historical conflicts. The use of advanced weaponry, heavy artillery, and explosives increases the scope of damage to natural landscapes. Military convoys and tanks leave deep scars on the land, compacting soil and destroying vegetation, while airstrikes devastate infrastructure that spills hazardous materials into the environment.

Moreover, the logistics of war—fuel depots, supply chains, and military camps—place additional strain on natural resources. Large-scale troop movements and the establishment of makeshift bases often lead to deforestation, soil erosion, and the overuse of local resources. The deliberate targeting of critical infrastructure, such as oil refineries, chemical plants, and energy facilities, amplifies environmental harm, releasing pollutants that affect ecosystems and human populations alike.

Modern warfare also introduces technological elements, such as drones and automated systems, which may seem less destructive but still contribute to environmental stress. For example, the energy demands of operating these systems, combined with the use of fuel-intensive machinery, add to the carbon footprint of war. The environmental costs of these operations are rarely accounted for in the aftermath, leaving affected regions to grapple with the consequences for years to come.

In summary, the immediate environmental cost of war is vast and multifaceted, touching every aspect of the natural world. Forests are razed, rivers are polluted, and the air becomes toxic—all within a matter of days or weeks of active combat. Modern military operations, with their reliance on high-intensity weaponry and infrastructure targeting, only accelerate these impacts, creating long-term challenges for recovery and sustainability. The Russia-Ukraine conflict stands as a stark reminder of how war not only devastates human lives but also leaves an indelible mark on the planet itself.

1.2 Ecosystems under Siege: The Hidden Victims of War

War spares no aspect of the natural world, and ecosystems, the intricate webs of life that sustain biodiversity, are among the most overlooked casualties. In the Russia-Ukraine conflict, the relentless bombings, shelling, and military operations have turned vibrant landscapes into battle-scarred terrains. The physical destruction of land and water systems is only the beginning; the long-term effects ripple through ecosystems, disrupting habitats, forcing wildlife migration, and threatening biodiversity.

Scars on the Natural Landscape

The impact of bombings and shelling on the natural environment is both immediate and devastating. Craters from explosive devices pockmark the land, uprooting vegetation and destroying topsoil essential for agriculture and natural regeneration. Once fertile fields are reduced to barren wastelands, and forests—home to countless species—are shredded by the sheer force of military operations. These scars are not merely superficial; they alter the very structure of the land, making recovery a slow and uncertain process.

Military activity often targets industrial zones and urban areas, but the collateral damage extends far beyond these focal points. Adjacent forests and grasslands become dumping grounds for debris and contaminants, compounding the environmental damage. The Donbas region, for example, known for its industrial significance and rich natural surroundings, has suffered extensive environmental degradation, with its landscapes now resembling desolate battlefields rather than thriving ecosystems.

Long-Term Effects on Soil and Water Systems

The immediate destruction from warfare is accompanied by lingering impacts on soil and water systems, which form the foundation of life for both humans and wildlife. The heavy machinery and explosives used in modern conflicts compact and erode the soil, stripping it of nutrients and rendering it unfit for agriculture or natural vegetation. Contaminants from military operations, including fuel spills, heavy metals, and chemicals, seep into the ground, poisoning the soil and water sources for decades.

Waterways are especially vulnerable, as the bombing of industrial facilities near rivers and lakes introduces hazardous substances into aquatic ecosystems. In Ukraine, the contamination of rivers like the Dnipro and its tributaries has disrupted not only human access to clean water but also the delicate balance of aquatic life. This pollution affects fish populations, aquatic plants, and the broader food chains that depend on these waterways. Restoring such systems requires not only cleaning but also the reestablishment of ecological balance, a process that can take years or even generations.

The Chain Reaction: Habitat Disruption, Wildlife Migration, and Biodiversity Loss

The destruction of landscapes and ecosystems sets off a chain reaction that impacts wildlife in profound ways. Habitats that once provided shelter and sustenance are obliterated, leaving species with nowhere to turn. Birds, mammals, and other animals are forced to migrate, often into unfamiliar and unsuitable environments where they struggle to survive. This displacement disrupts ecological networks, leading to declines in populations and, in some cases, the local extinction of species.

The loss of biodiversity is another critical consequence of ecosystem destruction. War zones often overlap with regions of significant ecological value, including forests, wetlands, and grasslands. As these areas are damaged, the species that depend on them are pushed to the brink. For example, Ukraine's steppe regions, known for their unique flora and fauna, have faced habitat destruction that threatens rare and endangered species. These losses are not just regional concerns; they have global implications for conservation and ecological stability.

Moreover, the disruption caused by war impacts the natural processes that ecosystems rely on to thrive. Pollinators like bees and butterflies may vanish from war-torn areas, affecting plant reproduction and agricultural productivity. Predators and prey are forced out of their established habitats, throwing off population balances and leading to cascading effects throughout the ecosystem.

Thus, the destruction of ecosystems during war often goes unnoticed in the immediate aftermath, overshadowed by human suffering and economic challenges. Yet these hidden victims of conflict bear the weight of damage that takes years to repair, if it can be repaired at all. The scars on the natural landscape, the contamination of soil and water systems, and the loss of habitats and biodiversity are stark reminders of the far-reaching consequences of modern warfare. Recognizing and addressing these impacts is crucial to understanding the true cost of war and the urgent need for global efforts to protect and restore our natural world.

1.3 The Air We Breathe: Pollution from Conflict Zones

Warfare not only devastates the land and water but also pollutes the very air we breathe, leaving long-lasting impacts on both human health and the environment. The Russia-Ukraine conflict has unleashed a barrage of airborne pollutants, from toxins released by explosions and chemical spills to smoke from burning infrastructure. The degradation of air quality in and around conflict zones poses immediate dangers to human populations while contributing to global atmospheric damage that accelerates climate change.

Airborne Toxins: Explosions, Chemical Spills, and Burning Infrastructure

Modern warfare generates a significant amount of air pollution, with explosions, fires, and industrial destruction releasing a cocktail of harmful substances into the atmosphere. Explosives and ammunition often contain chemicals that, when detonated, release toxic gases such as carbon monoxide, sulfur dioxide, and nitrogen oxides. In urban areas, the destruction of buildings and infrastructure produces fine particulate matter (PM2.5 and PM10), which remains suspended in the air and can be carried over long distances by wind.

Chemical spills from damaged industrial facilities and oil depots further exacerbate air pollution. When these structures are targeted or inadvertently destroyed during the conflict, hazardous chemicals, including volatile organic compounds (VOCs) and heavy metals, are released into the air. Burning oil fields, power plants, and storage tanks generate plumes of smoke laden with soot and toxic chemicals, creating a suffocating environment for nearby populations. In Ukraine, the targeting of energy infrastructure has resulted in widespread air pollution, with thick black smoke from burning fuel and power stations choking the skies over cities and rural areas alike.

Impact on Human Health and Regional Air Quality

The air pollution generated during conflicts poses immediate and long-term health risks to civilians and combatants alike. Fine particulate matter from explosions and fires penetrates deep into the lungs, leading to respiratory illnesses such as asthma, bronchitis, and chronic obstructive pulmonary disease (COPD). Prolonged exposure to these pollutants can also increase the risk of cardiovascular disease, lung cancer, and premature death.

Children, the elderly, and individuals with pre-existing health conditions are particularly vulnerable to the effects of polluted air. In conflict zones, access to healthcare is often limited, exacerbating the suffering caused by respiratory and cardiovascular issues. Civilians trapped in areas under heavy bombardment may face prolonged exposure to toxic air, with little recourse for relief. The psychological stress of living in such conditions compounds the physical health risks, creating a dire humanitarian crisis.

Beyond immediate health concerns, the deterioration of regional air quality has broader implications for neighboring countries. Airborne pollutants do not respect borders; they can travel hundreds or even thousands of miles, spreading the effects of war-induced air pollution far beyond the conflict zone. Neighboring regions may experience diminished air quality, acid rain, and other environmental consequences, underscoring the global reach of localized conflicts.

How Atmospheric Damage Contributes to Climate Change

The environmental impacts of war are not confined to the ground—they extend into the atmosphere, contributing to the global challenge of climate change. Burning fossil fuel infrastructure, such as oil refineries and storage tanks, releases massive amounts of carbon dioxide (CO_2), a key greenhouse gas, into the atmosphere. Fires caused by military operations also emit black carbon, a short-lived climate pollutant that absorbs sunlight and accelerates the melting of ice and snow when deposited on these surfaces.

The destruction of forests and vegetation, which act as natural carbon sinks, further exacerbates climate change by reducing the Earth's capacity to absorb CO_2. In the Russia-Ukraine war, the widespread loss of green cover due to bombings and military activity has contributed to increased atmospheric carbon levels. This double-edged impact—direct emissions from burning and the loss of carbon sequestration—intensifies the climate crisis.

Moreover, the release of toxic gases and particulate matter from military operations alters local and regional weather patterns. Air pollution can lead to the formation of smog and haze, reducing visibility and affecting sunlight penetration. Over time, these disruptions can impact agricultural productivity and natural ecosystems, adding to the already devastating effects of war.

In summary, the air pollution generated during the Russia-Ukraine conflict highlights the far-reaching environmental and humanitarian costs of war. From the toxic clouds that hang over battlefields to the long-term contribution to global warming, the degradation of air quality serves as a stark reminder of how warfare transcends borders and generations. Addressing the atmospheric impacts of conflict is critical to mitigating its broader consequences, both for the health of affected populations and for the planet's climate future.

1.4 Rivers in Ruin: The Water Crisis in War-Torn Areas

Water, the lifeblood of ecosystems and human civilization, is often a silent casualty of war. In the Russia-Ukraine conflict, rivers and water systems have faced unprecedented damage from bombings, chemical contamination, and infrastructure destruction. The impacts extend far beyond the immediate loss of water resources, affecting agriculture, drinking water supplies, and aquatic ecosystems. The damage to key rivers and lakes in Ukraine highlights the far-reaching consequences of war on one of the planet's most vital resources.

Destruction of Water Sources: Bombings and Chemical Contamination

Military operations in war zones often target or inadvertently destroy infrastructure critical to water systems, including dams, pipelines, and industrial facilities located near rivers. Explosions and shelling release debris, fuel, and hazardous chemicals into water bodies, contaminating them for years. In Ukraine, heavy fighting in industrial areas has led to the leakage of toxic substances such as oil, ammonia, and heavy metals into rivers and lakes, rendering them unsafe for consumption or agricultural use.

Bombings near reservoirs and water treatment plants have compounded the crisis. These facilities, essential for ensuring clean water supplies, are often damaged or destroyed, leaving millions without access to potable water. The deliberate targeting of such infrastructure by military forces exacerbates the water crisis, creating conditions for outbreaks of waterborne diseases and worsening the humanitarian situation.

Cascading Effects on Agriculture, Drinking Water, and Aquatic Ecosystems

The contamination and destruction of water sources have a cascading impact on multiple facets of life. For agriculture, polluted rivers and disrupted irrigation systems spell disaster for crops, leading to food shortages and economic instability. Farmers, unable to access clean water for irrigation, face reduced yields and the loss of livelihoods, while communities dependent on these agricultural systems endure heightened food insecurity.

For drinking water supplies, the damage is equally severe. Millions of civilians in war-torn areas rely on rivers and reservoirs for their daily water needs. Contamination with industrial chemicals and sewage makes these sources unsafe, forcing people to turn to alternative supplies, which are often scarce or expensive. This lack of access to clean drinking water not only increases the risk of disease but also contributes to the displacement of populations seeking better living conditions.

Aquatic ecosystems are profoundly affected by the pollution and physical destruction caused by war. Rivers, home to diverse species of fish, plants, and microorganisms, suffer from habitat disruption and toxic contamination. The delicate balance of these ecosystems is disrupted, leading to declines in biodiversity and long-term ecological damage. The ripple effects extend to the broader environment, as damaged aquatic systems affect the health of surrounding terrestrial ecosystems and communities.

Key Rivers and Lakes Impacted During the Russia-Ukraine Conflict

The Dnipro River, Ukraine's largest and most important waterway, has been a major casualty of the conflict. Stretching over 2,200 kilometers and supplying water to millions, the Dnipro has faced significant pollution from industrial zones and military activities. Bombings near its tributaries and chemical leaks have contaminated vast stretches, threatening the health of both aquatic ecosystems and human populations.

The Donets River, flowing through eastern Ukraine, has also been heavily impacted. Situated in a region with intense military activity, the Donets has been contaminated with industrial waste, fuel, and debris from bombings. Its once-thriving ecosystems have suffered, and the communities relying on it for drinking water and irrigation face an acute crisis.

Lakes and reservoirs, such as the Kakhovka Reservoir, have similarly experienced degradation. The destruction of surrounding infrastructure has led to reduced water quality and supply, while pollution and sedimentation have harmed aquatic life. These water bodies, crucial for agriculture and local economies, are now symbols of the environmental toll of war.

Thus, the water crisis in war-torn areas like Ukraine underscores the profound and lasting damage that conflicts inflict on vital natural resources. Rivers and lakes, essential for sustaining life, become battlegrounds in their own right, bearing scars that take decades to heal. Addressing the damage to these systems is not only a matter of environmental recovery but also a cornerstone of rebuilding communities and ensuring long-term stability in post-war regions.

1.5 Scorched Earth: The Destruction of Forests

Forests, often regarded as the lungs of the planet, are among the most vulnerable natural resources during armed conflicts. In the Russia-Ukraine war, these vital ecosystems have suffered extensively, becoming unintended casualties of relentless military operations. The destruction of forests not only disrupts local ecosystems but also contributes to global climate challenges by reducing the planet's ability to sequester carbon. Specific cases of deforestation and habitat loss in Ukraine highlight the environmental tragedy unfolding in the shadow of war.

Forests as Unintended Casualties of Military Operations

Military campaigns in forested regions result in widespread devastation, often as collateral damage. In the heat of battle, forests are obliterated by artillery fire, airstrikes, and ground combat. Missile strikes and shelling ignite wildfires, consuming vast tracts of woodland in a matter of hours. Heavy military equipment, such as tanks and armored vehicles, crush vegetation and compact soil, making reforestation efforts challenging.

Forests are also cleared for tactical reasons, such as to deny cover to opposing forces or to create clear lines of sight for military operations. Even when not directly targeted, forests become dumping grounds for debris, unexploded ordnance, and hazardous materials, further degrading their health. In Ukraine, regions like the Carpathians and forested areas near the Donbas have borne the brunt of these destructive activities, transforming once-thriving ecosystems into wastelands.

Loss of Carbon Sequestration and Its Global Climate Implications

The destruction of forests in conflict zones has implications that extend far beyond national borders. Forests play a crucial role in mitigating climate change by absorbing carbon dioxide from the atmosphere. When forests are destroyed, this carbon-sequestering capacity is lost, and the carbon stored in trees is released into the atmosphere as greenhouse gases, exacerbating global warming.

In Ukraine, the loss of forest cover due to military activity contributes to a vicious cycle of climate change. Burning forests release not only carbon dioxide but also black carbon, a short-lived climate pollutant that accelerates the melting of ice and snow in polar regions. The ripple effects of these emissions are felt globally, highlighting the interconnectedness of environmental crises.

Moreover, the destruction of forests reduces the planet's ability to recover from climate change. Reforestation efforts are hampered in conflict zones due to land degradation, contamination, and the presence of unexploded ordnance. This makes it difficult for ecosystems to regenerate naturally, leaving a long-term gap in the global fight against climate change.

Specific Cases of Deforestation and Habitat Loss in Ukraine

The Carpathian Mountains, home to some of Ukraine's most biodiverse forests, have faced significant threats due to the ongoing conflict. Military operations and the displacement of communities have led to deforestation in this region, endangering rare species and disrupting local ecosystems. The Carpathians, known for their role in regulating regional climate and water cycles, have suffered habitat loss that will take decades to recover.

The forested areas in eastern Ukraine, particularly in the Donbas region, have been heavily impacted by shelling and ground combat. These forests, already under pressure from industrial activity before the war, have seen extensive damage as a result of direct military actions. The loss of tree cover in these regions has disrupted local wildlife, forcing species to migrate or face extinction.

Another notable example is the steppe woodlands of southern Ukraine, which have been subjected to fires caused by missile strikes. These areas, characterized by their unique mix of grasslands and forests, have experienced habitat fragmentation and loss, impacting species that rely on these ecosystems for survival.

In summary, the destruction of forests during the Russia-Ukraine war is a stark reminder of the environmental costs of modern conflict. These ecosystems, essential for maintaining biodiversity and combating climate change, are being lost at an alarming rate. The global implications of deforestation in war zones demand urgent attention and action to mitigate damage, restore habitats, and prioritize the protection of forests in conflict-affected regions. The loss of these natural assets is not just a regional tragedy but a global challenge that underscores the need for a unified approach to environmental preservation in times of war.

1.6 Case Studies: Key Affected Areas in Ukraine

The Russia-Ukraine war has left profound environmental scars across the country, with certain regions bearing the brunt of its devastation. Each area tells a unique story of how war has compounded existing challenges, disrupted ecosystems, and created long-lasting environmental crises. This section explores three key affected areas: the industrially significant Donbas region, the vital Dnipro River Basin, and the already fragile Chornobyl Exclusion Zone. Together, these case studies illustrate the far-reaching environmental consequences of modern warfare.

Donbas Region: Environmental Degradation in an Industrial War Zone

The Donbas region, long known as Ukraine's industrial heartland, has become one of the most heavily impacted areas of the war. With its coal mines, steel plants, and chemical factories, the region was already facing significant environmental challenges before the conflict began. The ongoing war has turned this industrial zone into a battlefield, amplifying the environmental degradation to catastrophic levels.

The relentless shelling and bombardment have caused extensive damage to industrial facilities, leading to chemical spills, air pollution, and the release of hazardous substances into the soil and water. Coal mines, many of which are now abandoned or partially operational, are flooding, releasing toxic mine water into the surrounding environment. This has contaminated nearby rivers and aquifers, posing serious risks to both human health and aquatic ecosystems.

The destruction of infrastructure in the Donbas has also resulted in the release of heavy metals and other pollutants into the air and soil. The combination of industrial pollutants and wartime destruction has turned parts of the region into a toxic wasteland, making it one of the most environmentally hazardous zones in Ukraine.

Dnipro River Basin: Pollution and Ecosystem Disruption

The Dnipro River, Ukraine's most important waterway, has not escaped the environmental toll of the war. Stretching over 2,200 kilometers, the river supplies drinking water, supports agriculture, and sustains aquatic ecosystems. However, the conflict has severely disrupted its delicate balance.

Military activity near the Dnipro River Basin has led to widespread pollution, with industrial spills, fuel leaks, and debris contaminating the water. The destruction of dams and water infrastructure has further exacerbated the situation, causing flooding in some areas and water shortages in others. This has disrupted agricultural activities, which rely heavily on the river for irrigation, and has created a cascading effect on food security in the region.

Aquatic ecosystems in the Dnipro River Basin have also suffered immensely. Fish populations have declined due to contamination and habitat disruption, affecting local fisheries and the communities that depend on them. The pollution has also threatened the river's biodiversity, endangering plant and animal species that play crucial roles in maintaining ecological stability.

Chornobyl Exclusion Zone: The Compounded Impact of War on an Already Fragile Ecosystem

The Chornobyl Exclusion Zone, already one of the most fragile ecosystems on the planet, has faced new threats due to the ongoing conflict. As a site of the world's worst nuclear disaster, the zone had been slowly recovering over decades, with its unique ecosystem becoming a haven for wildlife. However, military activity in the area has disrupted this precarious recovery.

The movement of heavy military equipment through the exclusion zone has stirred up radioactive dust, spreading contamination to previously unaffected areas. Fires caused by shelling have further exacerbated the situation, releasing radioactive particles into the air and affecting both local and regional air quality. These disturbances have not only put military personnel at risk but have also reignited concerns about the spread of radiation to surrounding regions.

The conflict has also disrupted ongoing environmental monitoring and research in the zone. Scientists and conservationists who were working to study and rehabilitate the area have been forced to abandon their efforts, leaving the Chornobyl ecosystem vulnerable to unchecked damage. The war has compounded the zone's existing environmental challenges, turning it into yet another casualty of the conflict.

These case studies highlight the diverse and far-reaching environmental impacts of the Russia-Ukraine war. From the toxic industrial wasteland of the Donbas to the polluted waters of the Dnipro River and the fragile Chornobyl Exclusion Zone, each region underscores the urgency of addressing the environmental toll of modern warfare. These areas serve as stark reminders of the long-lasting scars that war leaves on the natural world, calling for immediate attention and action to mitigate their effects and support recovery efforts.

Interview with Dr. Elena Kovalchuk, Environmental Scientist in Conflict Zones

Dr. Elena Kovalchuk is an environmental scientist specializing in biodiversity and ecosystem restoration in conflict-affected areas. With over a decade of experience, she has worked extensively in Ukraine, assessing the environmental impacts of the ongoing conflict and developing strategies to mitigate its effects.

Q: Dr. Kovalchuk, thank you for speaking with us. Could you begin by describing how war impacts biodiversity and ecosystems in conflict zones like Ukraine?

A: War is devastating to ecosystems in ways that are both immediate and long-lasting. In Ukraine, we've seen forests destroyed by bombings and shelling, wetlands polluted with chemical runoff, and rivers contaminated by industrial spills from targeted facilities. The war doesn't just damage physical infrastructure—it also disrupts entire ecosystems. Wildlife is displaced as their habitats are destroyed or rendered uninhabitable, leading to shifts in local biodiversity. Migratory patterns are disrupted, and species that rely on specific environmental conditions may not survive.

For example, the Donbas region, known for its diverse ecosystems, has been heavily affected. Bomb craters have created physical barriers for wildlife, while chemical pollutants from damaged industrial sites have seeped into the soil and water, impacting both flora and fauna.

Q: What are some of the specific challenges you face in monitoring environmental damage in a conflict zone?

A: Monitoring environmental damage during an ongoing conflict is incredibly challenging for several reasons. First, access to affected areas is often limited due to safety concerns. Active combat zones are too dangerous to enter, and even regions that appear stable can be littered with unexploded ordnance, posing significant risks.

Second, there's the challenge of data collection. Many of the tools we typically use—such as ground surveys, drones, and remote sensing equipment—are difficult to deploy in war zones. Drones, for example, may be perceived as military equipment and could be targeted. We rely heavily on satellite imagery and remote data collection, but this often provides only a partial picture.

Lastly, there's the issue of time. Environmental monitoring typically requires longitudinal studies to understand changes over months or years. In a war zone, conditions change rapidly, making it difficult to establish baselines or track long-term trends. The urgency of immediate humanitarian crises often overshadows the need for ecological monitoring, further complicating our work.

Q: Can you share any specific examples of environmental damage you've documented in Ukraine?

A: One stark example is the pollution of the Dnipro River Basin. Industrial facilities along the river have been damaged by shelling, releasing toxic chemicals into the water. This not only affects aquatic ecosystems but also poses a significant risk to human communities that rely on the river for drinking water and agriculture.

Another case is the forests in the Donetsk region, where military activity has led to large-scale deforestation. Fires caused by bombings have destroyed hectares of woodland, releasing stored carbon into the atmosphere and contributing to climate change. These forests were home to several protected species, many of which are now at risk due to habitat loss.

Additionally, the Chornobyl Exclusion Zone, already fragile due to its history, has seen renewed stress. Military movements and shelling in the area have disrupted soil and vegetation, potentially re-releasing radioactive particles into the environment.

Q: What strategies can be used to mitigate these environmental impacts during and after a conflict?

A: During a conflict, mitigation efforts are understandably limited, but there are still steps that can be taken. For example, creating buffer zones around eco-sensitive areas, like wetlands or forests, can reduce the likelihood of direct damage. Additionally, prioritizing the protection of critical infrastructure—such as water treatment plants and industrial facilities—can prevent cascading environmental disasters.

Post-conflict, the focus shifts to restoration. Reforestation efforts are crucial for areas that have experienced deforestation, but this must be done thoughtfully to avoid introducing invasive species. Contaminated soil and water require extensive cleanup, often involving phytoremediation—using plants to absorb and neutralize pollutants.

International collaboration is key. Ukraine has already received support from environmental NGOs and international organizations, which have provided funding and expertise for recovery efforts. However, what's most needed is a long-term commitment to integrating environmental restoration into post-war reconstruction plans.

Q: What gives you hope as you work in these challenging conditions?

A: Despite the devastation, nature has a remarkable ability to heal—if given the chance. I've seen wetlands slowly regenerate after being polluted, and forests beginning to reclaim areas that were once battlegrounds. The resilience of ecosystems mirrors the resilience of the communities living within them.

What gives me the most hope is the growing global awareness of the environmental costs of war. More nations and organizations are recognizing that environmental protection isn't just a secondary issue—it's fundamental to rebuilding lives and ensuring long-term stability. The willingness of people to come together, share knowledge, and work toward recovery inspires me every day.

Q: Lastly, what message would you like to share with the global community?

A: My message is simple: environmental protection must be prioritized, even during conflict. The damage we allow today will affect generations to come—not just in Ukraine but around the world. Ecosystems are interconnected, and their health is tied to ours. If we invest in safeguarding and restoring the environment now, we're investing in the future of humanity.

Dr. Elena Kovalchuk's insights remind us that the environment is an often-overlooked casualty of war. Her work in Ukraine highlights the critical need for global attention to the ecological impacts of conflict and the importance of integrating environmental recovery into peacebuilding efforts.

1.7 The Lasting Legacy of Environmental Damage

The environmental damage caused by war is not confined to the duration of the conflict; it leaves an enduring legacy that reshapes ecosystems for decades. The Russia-Ukraine war is no exception, with its devastating impact on the natural world likely to be felt long after the fighting ceases. The scars left on the land, water, and air pose significant challenges for restoration and carry ripple effects that extend far beyond Ukraine's borders. Understanding these long-term consequences is crucial to addressing the full scope of wartime destruction.

How Wartime Destruction Shapes Ecosystems for Decades

The damage inflicted on ecosystems during war often triggers a chain reaction of environmental degradation that can persist for generations. Bomb craters and artillery strikes disrupt the soil structure, making it less fertile and harder for vegetation to grow. Contaminants such as heavy metals, oil, and toxic chemicals seep into the ground and water, creating conditions that are inhospitable for plant and animal life. In many cases, these pollutants remain in the environment for decades, causing chronic harm to ecosystems.

Wildlife is another casualty of war, with habitats destroyed and species forced to migrate or face extinction. The loss of biodiversity can destabilize entire ecosystems, reducing their resilience to other stresses, such as climate change or human development. The once-thriving ecosystems of Ukraine's forests, rivers, and wetlands are now fragmented and at risk of collapse, a process that will take years to reverse, if it can be reversed at all.

Challenges in Rebuilding and Restoring Natural Landscapes

Restoring war-damaged environments is a daunting and often underfunded task. The process is complicated by the presence of unexploded ordnance, landmines, and military debris, which pose risks to both humans and wildlife. These remnants of war must be safely removed before any meaningful restoration efforts can begin, a process that can take years and require significant resources.

Soil contamination is another major challenge. Rehabilitating polluted land often involves costly and labor-intensive techniques, such as removing and treating the topsoil or introducing bio-remediation methods. In areas where water systems have been polluted, restoration efforts are equally complex, involving the cleanup of contaminants and the re-establishment of aquatic ecosystems.

The lack of resources and infrastructure in post-war regions often delays these efforts. Governments and international organizations are frequently preoccupied with rebuilding cities and addressing humanitarian needs, leaving environmental recovery as a secondary priority. However, the longer restoration is delayed, the more difficult and expensive it becomes, compounding the long-term impact on ecosystems and communities.

The Global Ripple Effects of Local Environmental Crises

The environmental damage caused by war does not remain confined to the affected region. Local crises can have far-reaching implications, influencing global ecosystems, economies, and even geopolitics. For example, the pollution of Ukraine's rivers, such as the Dnipro, not only impacts local water supplies but also affects international water systems connected to the Black Sea, disrupting marine ecosystems and fisheries in neighboring countries.

Similarly, the destruction of forests in Ukraine, a key carbon sink in the region, contributes to the global climate crisis by increasing greenhouse gas emissions and reducing the planet's capacity to absorb carbon dioxide. The loss of biodiversity in one region can also have cascading effects, weakening ecological networks that span across borders.

Moreover, environmental damage in conflict zones often triggers migration, as people are forced to leave areas that can no longer sustain life. This displacement creates additional pressure on neighboring countries and ecosystems, further exacerbating global environmental and humanitarian challenges. The interconnected nature of today's world means that the environmental costs of war in one country are ultimately shared by all.

Thus, the lasting legacy of environmental damage from the Russia-Ukraine war is a stark reminder of the enduring consequences of conflict. From ecosystems disrupted for decades to the challenges of restoration and the global ripple effects of local crises, the environmental toll of war extends far beyond the battlefield. Addressing this legacy requires a concerted effort from governments, international organizations, and local communities to prioritize environmental recovery as a key component of post-war rebuilding. Only by recognizing and addressing these long-term impacts can we hope to mitigate the lasting scars of war on the natural world.

Conclusion: The Battlefield and Nature

The Russia-Ukraine war is a stark reminder of the profound and lasting damage that modern conflicts inflict on the environment. From scorched forests and polluted rivers to toxic air and disrupted ecosystems, the natural world becomes an unacknowledged casualty of human aggression. These impacts are not confined to the immediate theater of war but ripple outward, leaving scars that last for decades and consequences that extend far beyond national borders.

This chapter has explored how warfare reshapes landscapes, disrupts biodiversity, and accelerates global environmental crises. The destruction of forests and ecosystems reduces the planet's ability to mitigate climate change, while the contamination of water and air threatens both human health and ecological stability. Case studies such as the Donbas region, the Dnipro River Basin, and the Chornobyl Exclusion Zone highlight the diverse and far-reaching nature of these challenges, painting a grim picture of war's toll on nature.

The legacy of environmental damage left in the wake of this conflict underscores the urgent need for a holistic approach to understanding and addressing the true costs of war. Recovery efforts must prioritize not only human rebuilding but also ecological restoration, recognizing that the health of the environment is intrinsically tied to the well-being of societies.

As the world watches the ongoing conflict, it is crucial to acknowledge the silent suffering of the natural world. The scars of war on the Earth are reminders of the interconnectedness of humanity and the environment. Addressing these damages requires global cooperation, innovative solutions, and a commitment to safeguarding the ecosystems that sustain us all. Only then can we begin to heal the wounds inflicted on both people and the planet.

Chapter 2: Toxic Legacies

Every war leaves behind scars, but some are far more enduring than others. The Russia-Ukraine conflict has created a toxic legacy that will linger long after the fighting ceases, impacting the land, water, and lives of those caught in its path. The relentless use of chemical weapons, the destruction of industrial facilities, and the spillage of oil and hazardous materials have unleashed pollutants that permeate the environment, transforming once-thriving regions into contaminated wastelands. These toxic remnants of war do not merely disrupt ecosystems in the short term—they create long-lasting challenges for agriculture, biodiversity, and public health.

The invisible yet insidious effects of environmental degradation seep into the soil, flow through water systems, and infiltrate the air, threatening the health of civilians and soldiers alike. In a world increasingly interconnected by ecological and economic dependencies, the toxic legacies of war extend far beyond the borders of Ukraine, creating global implications that demand urgent attention. This chapter delves into the persistent environmental and human costs of war, examining the lasting scars left by toxic pollutants and the monumental task of rebuilding amid the remnants of destruction.

2.1 Long-Term Environmental Fallout from Chemical Weapons and Industrial Damage

The environmental consequences of war are not confined to the immediate destruction of battlefields. Modern conflicts, like the Russia-Ukraine war, leave behind toxic legacies that persist for years, if not decades. Chemical residues, hazardous waste, oil spills, and fires caused by the destruction of industrial facilities contribute to widespread contamination. These pollutants infiltrate the land, water, and air, creating a cascade of environmental and health crises that extend far beyond the war's end.

Persistent Pollutants: The Enduring Presence of Chemical Residues

The use of chemical weapons and the destruction of industrial infrastructure release a cocktail of toxic substances into the environment. Explosives and ammunition often contain heavy metals like lead, mercury, and arsenic, which seep into the soil and water after detonation. These pollutants can remain in the environment for decades, poisoning agricultural lands and endangering ecosystems.

For example, chemical residues from weapons such as phosphorus bombs or thermobaric munitions linger in the soil and air, causing harm to both humans and wildlife. These substances can disrupt plant growth, contaminate food chains, and alter the natural composition of ecosystems. The enduring presence of such pollutants ensures that even areas no longer involved in active combat continue to suffer from reduced productivity and biodiversity.

Industrial Destruction: A Source of Hazardous Waste and Emissions

Industrial facilities often become strategic targets in modern warfare, and their destruction has devastating environmental consequences. When factories, chemical plants, or refineries are bombed, they release hazardous waste into the surrounding environment. Chemicals stored in tanks or pipelines spill onto the ground, infiltrating water systems and contaminating the soil.

In Ukraine, the targeting of industrial zones, particularly in the Donbas region, has turned these areas into toxic hotspots. The release of ammonia, chlorine, and other industrial chemicals from damaged facilities poses a severe threat to nearby communities. In addition to localized contamination, the gases and particulate matter released during these events contribute to air pollution, affecting regions far from the original site of destruction.

The cleanup of hazardous waste from industrial damage is a monumental challenge, requiring extensive resources and expertise. Without immediate intervention, these pollutants spread, worsening their impact and complicating future restoration efforts.

Oil Spills as Ecological Disasters: Spread, Impact, and Challenges in Containment

Oil spills caused by the targeting of fuel depots, pipelines, and storage facilities are another significant environmental legacy of war. The deliberate or accidental release of oil into the environment creates ecological disasters, contaminating soil, water, and air. Once spilled, oil spreads rapidly, affecting large areas and penetrating ecosystems deeply.

In Ukraine, the destruction of oil storage facilities has led to extensive contamination of rivers, lakes, and agricultural lands. The toxic components of oil—such as polycyclic aromatic hydrocarbons (PAHs)—are harmful to aquatic life, plants, and animals. They reduce oxygen levels in water, suffocate fish, and destroy habitats, disrupting entire food chains.

Containment and cleanup efforts for oil spills during conflict are fraught with challenges. Limited access to war-torn regions, ongoing military activity, and inadequate resources often prevent timely responses. As a result, the environmental damage escalates, leaving long-term ecological consequences that are difficult to reverse.

Toxic Fires: Burning Factories and Their Impact on Air and Soil

The destruction of factories, fuel depots, and storage facilities often results in massive fires that release a toxic mix of chemicals into the air. These fires emit particulate matter, carbon dioxide, sulfur dioxide, and other harmful pollutants, creating hazardous air quality for those living nearby. The black smoke from burning factories and oil facilities carries soot and toxic residues that settle on the ground, contaminating soil and water.

In conflict zones like Ukraine, these fires exacerbate already dire environmental conditions. The air becomes laden with pollutants that cause respiratory illnesses, while the soil becomes infertile due to the deposition of toxic ash and residues. Fires in chemical plants are particularly dangerous, as they release volatile organic compounds (VOCs) and other harmful substances that affect both local and regional ecosystems.

The long-term impact of such fires includes soil degradation, the death of microorganisms essential for plant growth, and the disruption of agricultural activities. These conditions make it nearly impossible for affected areas to recover without significant intervention, leaving communities struggling with environmental and economic losses.

The long-term environmental fallout from chemical weapons, industrial damage, oil spills, and toxic fires underscores the far-reaching consequences of modern warfare. These pollutants not only devastate the immediate environment but also create persistent challenges that hinder recovery efforts and pose risks to future generations. Addressing these toxic legacies requires immediate action, international cooperation, and innovative approaches to mitigate and restore the affected regions. Without such efforts, the environmental scars of war will remain a permanent reminder of its destructive power.

Interview with Dr. Andriy Melnyk, Chemical Engineer and Industrial Safety Expert

Dr. Andriy Melnyk is a chemical engineer with over 20 years of experience in industrial safety and environmental restoration. He has worked extensively in war-affected regions, focusing on mitigating the long-term impacts of chemical spills and industrial damage caused by armed conflict.

Q: Dr. Melnyk, thank you for joining us. Can you start by explaining how industrial and chemical spills during conflicts like the Russia-Ukraine war impact the environment?

A: Conflicts involving industrial areas create a perfect storm for environmental damage. When industrial facilities, such as chemical plants, oil refineries, or storage depots, are targeted or caught in the crossfire, they release hazardous substances into the surrounding environment. These substances seep into the soil, contaminate groundwater, and often make their way into rivers and lakes, causing widespread ecological harm.

For instance, when a fertilizer plant is bombed, the nitrogen and ammonia compounds released can drastically alter soil chemistry and water quality, leading to algal blooms that suffocate aquatic ecosystems. Similarly, oil spills from pipelines or storage tanks spread hydrocarbons that poison soil and water, harming plants, animals, and humans.

The damage doesn't end there. These pollutants can persist for years or even decades, making recovery an incredibly complex and resource-intensive process. The environmental legacy of such incidents is often felt long after the conflict ends.

Q: What are the key challenges in restoring areas contaminated by war-related industrial and chemical spills?

A: Restoring contaminated areas after a conflict is an uphill battle for several reasons. First, there's the issue of **identifying the extent of contamination**. In conflict zones, access to affected areas is often limited, and damage assessments are delayed. By the time teams can conduct surveys, pollutants may have spread far beyond their original site, complicating cleanup efforts.

Second, **resource limitations** are a significant hurdle. Effective remediation requires advanced technologies, skilled personnel, and substantial funding—resources that are often scarce in post-war environments. Governments are usually focused on rebuilding infrastructure and addressing humanitarian needs, so environmental restoration takes a backseat.

Third, the **complexity of contaminants** poses a unique challenge. War-related spills often involve a cocktail of chemicals, each requiring specific remediation strategies. For example:

- Heavy metals, like lead and mercury, require excavation and containment.
- Hydrocarbons demand bioremediation or thermal treatments to break down the pollutants.
- Persistent organic pollutants (POPs) often need high-tech solutions like incineration or chemical neutralization.

Finally, there's the issue of **long-term monitoring**. Even after initial cleanup efforts, the risk of pollutants leaching into water sources or spreading through soil remains. Continuous monitoring is critical to ensuring that the environment and local communities are safe.

Q: Can you walk us through the steps involved in rehabilitating such contaminated zones?

A: Certainly. Restoring contaminated areas typically involves several stages, each tailored to the specific nature of the damage:

1. **Assessment and Mapping**

The first step is to conduct a thorough assessment of the affected site. This includes soil and water sampling, air quality tests, and geospatial mapping to determine the extent of contamination.

Advanced tools like satellite imagery and ground-penetrating radar can provide invaluable data for understanding the scope of the problem.

1. **Containment**

Immediate containment measures are crucial to prevent further spread of contaminants. This may involve installing barriers, such as geomembranes, to isolate polluted soil or using booms to contain oil spills in water bodies.

1. **Remediation**

Once containment is in place, the focus shifts to remediation:

✓ **Bioremediation:** Using microorganisms to break down hydrocarbons and other organic pollutants.

✓ **Phytoremediation:** Employing plants to absorb heavy metals and stabilize soil.

✓ **Chemical Treatment:** Neutralizing toxic substances through chemical reactions.

✓ **Excavation and Removal:** For severe contamination, affected soil or water may need to be removed entirely and treated off-site.

1. **Rehabilitation**

After contaminants are addressed, efforts turn to rehabilitating the ecosystem. Reforestation, soil replenishment, and reintroduction of native species are common practices to restore biodiversity and ecological balance.

1. **Monitoring and Maintenance**

Post-remediation, long-term monitoring ensures that the site remains stable. Sensors, drones, and IoT technologies can track environmental parameters like water quality and soil health, providing early warnings of any recurring issues.

Q: Have you observed any notable examples of successful restoration efforts in war-torn areas?

A: Yes, there are a few examples that stand out. One is the rehabilitation of areas affected by the conflict in the Donbas region of Ukraine. In the aftermath of heavy industrial damage, international NGOs collaborated with local agencies to contain pollutants from bombed chemical plants. Using bioremediation techniques, they successfully restored portions of polluted wetlands, enabling the gradual return of native flora and fauna.

Another example is from the Syrian conflict, where oil spills from sabotaged facilities had devastated surrounding farmland. Cleanup efforts included deploying bio-absorbent materials to soak up hydrocarbons and using phytoremediation to detoxify the soil. These methods, though time-intensive, demonstrated the potential for recovery even in severely affected areas.

Q: What role does international cooperation play in these restoration efforts?

A: International cooperation is absolutely vital. No single country has the resources or expertise to tackle the multifaceted challenges of war-related contamination on its own. International organizations provide funding, technical expertise, and advanced technologies that might not be available locally.

For instance, the United Nations Environment Programme (UNEP) has been instrumental in coordinating environmental recovery in post-conflict regions. Their teams bring specialized knowledge and tools, such as portable laboratories for on-site chemical analysis. Similarly, collaborations with private companies and universities have accelerated the development and deployment of innovative remediation techniques.

However, true success requires sustained commitment. Short-term funding and sporadic interventions won't suffice for long-term recovery. Governments, NGOs, and international bodies must work together to create comprehensive, multi-year plans that address both immediate and future challenges.

Q: What message would you like to convey to policymakers and the global community regarding the toxic legacies of war?

A: My message is that the environment cannot be treated as collateral damage. The toxic legacies of war are not just ecological—they are human. Contaminated water, poisoned soil, and polluted air affect the health and livelihoods of millions. Addressing these issues is not a luxury; it's a necessity for sustainable recovery.

Policymakers must prioritize environmental restoration as an integral part of post-war reconstruction, allocating resources and fostering international partnerships to make it happen. The global community must also recognize its shared responsibility—pollutants don't respect borders, and their impacts are felt far beyond the immediate conflict zone.

Dr. Andriy Melnyk's insights reveal the complexities of addressing the toxic legacies of war. His work underscores the need for a proactive and collaborative approach to restoring contaminated environments, ensuring that communities and ecosystems alike can heal and thrive once again.

2.2 Contaminated Water and Soil: A Crisis for Agriculture and Biodiversity

Water and soil are the foundational elements of life, supporting agriculture, ecosystems, and human survival. In the wake of the Russia-Ukraine war, these vital resources have suffered extensive contamination from chemical spills, industrial destruction, and military activity. The resulting crisis has created a chain reaction of environmental damage, threatening food security, biodiversity, and public health. The contamination of water and soil not only disrupts local ecosystems but also sets off a cascade of long-term consequences that extend far beyond the immediate conflict zones.

The Chain Reaction: How Chemicals Seep into Soil and Water

The environmental damage caused by war often begins with the release of toxic substances into the soil and water. Explosives, ammunition, and destroyed industrial facilities release heavy metals, chemicals, and oils that infiltrate the ground and waterways. When these pollutants enter the soil, they bind with minerals, altering its natural composition and making it unsuitable for plant growth. Over time, rainfall and irrigation wash these contaminants deeper into the ground or carry them into nearby rivers, lakes, and reservoirs, spreading the pollution further.

In Ukraine, heavy military activity in industrial regions has created hotspots of contamination where the soil and water are saturated with toxins. These chemicals disrupt the natural nutrient cycles in the soil, impairing its fertility and making it difficult for vegetation to recover. The ripple effects extend to aquatic ecosystems, where pollutants accumulate, poisoning fish, plants, and other organisms.

Loss of Arable Land: Implications for Agriculture and Food Security

One of the most immediate consequences of soil contamination is the loss of arable land. Fertile agricultural lands, vital for sustaining local communities and economies, are rendered barren when pollutants destroy the natural balance of nutrients and microorganisms. Crops grown in contaminated soil absorb harmful substances, making them unsafe for consumption and reducing yields.

In Ukraine, known for its rich agricultural lands, the contamination of soil has significantly impacted food production. Fields that once produced wheat, corn, and sunflowers now struggle to sustain growth, threatening both local and global food supplies. Ukraine's role as a major exporter of grains means that this loss of productivity affects international food security, driving up prices and creating shortages in dependent countries.

The long-term degradation of soil fertility requires extensive rehabilitation efforts, including soil replacement, bio-remediation, and reforestation, all of which are costly and time-consuming. Without immediate action, the loss of arable land could lead to prolonged food crises in Ukraine and beyond.

Polluted Waterways: Impact on Fish Populations and Aquatic Ecosystems

Rivers, lakes, and reservoirs in war zones are particularly vulnerable to pollution, as they serve as both natural habitats and sources of drinking water and irrigation. The release of chemicals, heavy metals, and oil into these waterways disrupts aquatic ecosystems, leading to declines in fish populations and the collapse of food chains. Pollutants reduce oxygen levels in the water, suffocating fish and other aquatic life, while sedimentation caused by military activity destroys spawning grounds.

In Ukraine, key waterways such as the Dnipro River have been heavily impacted by chemical spills from industrial facilities damaged during the conflict. The contamination has led to the disappearance of native fish species, affecting local fisheries and communities that rely on them for sustenance and livelihoods. The degradation of aquatic ecosystems also has a cascading effect on biodiversity, as species that depend on water sources are forced to migrate or perish.

Restoring polluted waterways is a significant challenge, requiring not only the removal of contaminants but also the re-establishment of natural ecological processes. Without such efforts, the long-term health of these ecosystems remains at risk.

Biodiversity Under Siege: The Ripple Effects on Flora and Fauna

The contamination of water and soil creates a hostile environment for plants and animals, disrupting ecosystems and threatening biodiversity. Toxic substances in the soil hinder plant growth, reducing the availability of food and shelter for herbivores. As plant populations decline, so do the animals that rely on them, creating a domino effect throughout the food chain.

Wildlife in Ukraine, particularly in conflict-affected regions such as the Donbas and the Carpathians, has faced severe challenges due to habitat destruction and contamination. Endangered species are at heightened risk as their already limited habitats are polluted or destroyed. The migration of species seeking safer environments further disrupts ecological balances, leading to overpopulation in some areas and depopulation in others.

The loss of biodiversity has long-term implications for ecosystem stability and resilience. Each species plays a unique role in maintaining ecological health, and their disappearance can lead to the collapse of entire ecosystems. Efforts to protect and restore biodiversity in war-torn regions require addressing the root causes of contamination and ensuring the safe rehabilitation of affected habitats.

Groundwater Contamination: Hidden Threats to Human and Ecological Health

Groundwater, a critical source of drinking water for millions, is particularly vulnerable to contamination during conflicts. Pollutants from damaged industrial facilities, leaking pipelines, and military activity seep into the ground and infiltrate aquifers. Unlike surface water, groundwater contamination is difficult to detect and can remain hidden for years, creating long-term health and environmental risks.

In Ukraine, the contamination of groundwater has created significant challenges for rural communities that rely on wells for drinking water. Chemicals such as nitrates, heavy metals, and hydrocarbons have been found in underground water supplies, posing severe health risks to humans and animals. The presence of toxic substances in groundwater also impacts agriculture, as contaminated water used for irrigation introduces pollutants into crops and soil.

Addressing groundwater contamination requires advanced technologies and significant investment, as well as the establishment of monitoring systems to detect and manage pollutants. Without these measures, the hidden threats posed by groundwater contamination will continue to jeopardize public health and ecological stability.

The contamination of water and soil in conflict zones like Ukraine represents a critical crisis with far-reaching implications. From the loss of fertile land and polluted waterways to the threats posed to biodiversity and human health, the ripple effects of environmental degradation are profound. Addressing these challenges requires immediate action to contain pollutants, rehabilitate ecosystems, and prioritize sustainable recovery efforts. The restoration of water and soil is not only essential for rebuilding war-torn regions but also for safeguarding the future of agriculture, biodiversity, and public health on a global scale.

2.3 Health Impacts of Environmental Degradation on Civilians and Soldiers

The environmental degradation caused by war leaves a profound impact on human health, affecting both civilians and soldiers. The Russia-Ukraine conflict, with its extensive contamination of air, water, and soil, has created a health crisis that compounds the already dire humanitarian situation. From immediate health risks to long-term conditions, the consequences of toxic environments are far-reaching, disproportionately affecting vulnerable populations and those directly involved in combat. This section examines how exposure to degraded environments undermines physical and psychological well-being, creating challenges that endure long after the conflict ends.

Immediate Health Risks: Respiratory Illnesses and Chemical Exposure

One of the most immediate health impacts of environmental degradation during war is the prevalence of respiratory illnesses. Airborne pollutants from explosions, chemical spills, and fires—such as fine particulate matter (PM2.5), toxic gases, and soot—pose serious risks to respiratory health. Civilians living near conflict zones are often exposed to thick, polluted air for extended periods, leading to conditions such as asthma, bronchitis, and acute respiratory distress.

In addition to air pollution, direct exposure to chemicals released from damaged industrial facilities and munitions creates a host of acute health risks. Contact with substances such as ammonia, chlorine, and heavy metals can result in burns, poisoning, and organ failure. Contaminated water supplies further exacerbate the problem, causing gastrointestinal illnesses and infections, particularly among populations with limited access to clean drinking water and medical care.

Chronic Health Conditions: The Long-Term Consequences of Toxic Environments

The long-term health consequences of exposure to toxic environments are often even more severe. Pollutants that persist in the soil, water, and air can lead to chronic illnesses that manifest years after the initial exposure. Heavy metals like lead and mercury, commonly found in contaminated soil and water, can accumulate in the body, causing neurological disorders, kidney damage, and developmental issues in children.

Prolonged exposure to polluted air increases the risk of cardiovascular diseases, lung cancer, and other chronic respiratory conditions. In agricultural regions where crops and livestock are contaminated by toxic substances, the consumption of food and water becomes a slow poison for local populations. Cancer clusters and developmental abnormalities often emerge as grim reminders of the toxic legacy of war, affecting entire generations.

Vulnerable Populations: Children, the Elderly, and Pregnant Women in Conflict Zones

Environmental degradation disproportionately affects the most vulnerable members of society—children, the elderly, and pregnant women. Children, with their developing immune and respiratory systems, are particularly susceptible to the harmful effects of air and water pollution. Exposure to toxic substances can lead to stunted growth, cognitive impairments, and increased susceptibility to infections.

The elderly, who often have pre-existing health conditions, face heightened risks from respiratory and cardiovascular issues caused by polluted air. Limited mobility and access to healthcare further exacerbate their vulnerability. Pregnant women exposed to contaminated water, air, or soil are at risk of complications such as premature births, low birth weights, and birth defects. The effects on unborn children can have lifelong consequences, compounding the generational toll of environmental degradation.

Psychological Effects: Living Amidst Environmental Degradation

The psychological toll of living in degraded environments is an often-overlooked aspect of war's impact on health. The constant exposure to toxic surroundings, coupled with the destruction of homes and communities, creates a sense of hopelessness and despair. Anxiety, depression, and post-traumatic stress disorder (PTSD) are common among civilians living in conflict zones where environmental degradation is severe.

The loss of access to clean water, arable land, and healthy ecosystems disrupts daily life, creating a sense of insecurity and helplessness. People forced to drink contaminated water or breathe polluted air experience chronic stress, which can exacerbate existing mental health conditions and increase the risk of physical illnesses. The psychological impact is particularly pronounced among children, who may struggle to understand the long-term consequences of their toxic surroundings but still experience the stress and trauma of living in such environments.

Military Exposure: The Impact of Contaminated Battlefields on Soldiers' Health

Soldiers deployed in conflict zones are exposed to environmental hazards that can have serious health implications. Prolonged exposure to polluted air from explosions and fires, contaminated water sources, and toxic substances on battlefields can lead to both immediate and chronic health issues. Soldiers often inhale fine particulates and toxic fumes, resulting in respiratory conditions such as asthma and chronic obstructive pulmonary disease (COPD).

The use of depleted uranium in munitions or the presence of hazardous materials on the battlefield can increase the risk of cancer and other long-term illnesses for soldiers. The psychological stress of operating in environmentally degraded conditions adds another layer of complexity to their health challenges, contributing to mental health issues such as PTSD and anxiety.

In addition to physical ailments, military personnel often face limited access to medical care during and after deployment, making it difficult to address the health impacts of their exposure. These issues not only affect the soldiers themselves but also place a burden on healthcare systems and families, creating a ripple effect that extends far beyond the battlefield.

The health impacts of environmental degradation during the Russia-Ukraine war highlight the devastating human cost of toxic environments. From respiratory illnesses and chronic diseases to the disproportionate suffering of vulnerable populations and soldiers, the toll on physical and mental health is immense. Addressing these health challenges requires comprehensive strategies that include environmental cleanup, access to healthcare, and support for those affected. Without such measures, the toxic legacy of this war will continue to harm the health and well-being of civilians and soldiers for generations to come.

2.4 Economic and Social Costs of Toxic Legacies

The toxic legacies of war extend beyond environmental and health impacts, significantly influencing economic stability and social cohesion. The Russia-Ukraine war has left behind a trail of environmental degradation that imposes heavy financial burdens on governments and communities, disrupts livelihoods, and deepens social inequalities. The challenges of addressing these toxic legacies are immense, as they require resources, infrastructure, and international cooperation in a context where recovery is already strained. This section explores the economic and social costs of environmental destruction in war-torn regions, emphasizing the interconnected nature of these crises.

The Economic Burden of Environmental Cleanup and Healthcare

The financial cost of cleaning up environmental damage caused by war is staggering. Decontaminating polluted soil, rehabilitating waterways, and managing hazardous waste require significant investment in technology, expertise, and labor. In Ukraine, the destruction of industrial facilities, chemical plants, and oil depots has resulted in widespread contamination that will take years, if not decades, to address. The cleanup of these sites involves not only the removal of toxic substances but also the restoration of ecosystems, reforestation, and the rebuilding of damaged infrastructure.

Governments in war-affected regions often struggle to prioritize environmental restoration amid competing demands for housing, healthcare, and infrastructure rebuilding. The cost of addressing these toxic legacies can divert funds from other critical areas, slowing down the overall recovery process. For example, the remediation of a single heavily polluted industrial site can cost millions of dollars, leaving little room in national budgets for other essential services.

Healthcare systems also bear a significant financial burden due to the health impacts of environmental degradation. Treating chronic illnesses, respiratory conditions, and waterborne diseases caused by pollution requires long-term medical care and resources. This increased demand strains healthcare infrastructure, especially in conflict zones where facilities may already be damaged or overwhelmed. The economic toll on families forced to seek medical treatment further exacerbates financial insecurity, deepening the cycle of poverty and hardship.

The Displacement of Communities Due to Contaminated Land

Environmental contamination often renders large areas of land uninhabitable, forcing communities to abandon their homes and relocate. In Ukraine, areas near industrial zones, rivers, and battlefields have become toxic hotspots, with residents fleeing the long-term dangers posed by polluted soil and water. Displacement disrupts social structures, severs community ties, and creates additional pressure on neighboring regions and urban centers that must absorb displaced populations.

The forced migration of communities due to environmental degradation also raises issues of land ownership and resettlement. Families who lose their homes and land face legal and logistical challenges in reclaiming or rebuilding their lives elsewhere. This displacement can lead to long-term instability, as governments and organizations struggle to provide adequate housing, infrastructure, and support for those affected.

The psychological toll on displaced communities cannot be overstated. Losing one's home and livelihood due to contamination adds to the trauma of living through war, creating a population that is not only physically but also emotionally displaced. These challenges underscore the need for comprehensive policies that address both the environmental and social dimensions of displacement.

Loss of Livelihoods Tied to Agriculture, Fishing, and Natural Resources

The contamination of land and water directly impacts livelihoods tied to natural resources, particularly in rural areas where agriculture and fishing are the primary means of income. In Ukraine, the destruction of fertile farmland and the pollution of rivers like the Dnipro have left farmers and fishers struggling to sustain their livelihoods. Crops grown in contaminated soil are unsafe for consumption or export, while polluted waterways decimate fish populations, rendering fishing industries unviable.

The loss of these livelihoods creates a ripple effect, impacting local economies and exacerbating food insecurity. Communities that depend on agriculture and fishing for income face financial ruin, leading to increased poverty and unemployment. The inability to access clean land and water for productive use stalls recovery efforts, perpetuating economic instability in the affected regions.

Rebuilding these industries is a complex and costly process. It involves not only cleaning up the environment but also restoring infrastructure, such as irrigation systems and fish hatcheries, and providing financial and technical support to affected communities. Without targeted interventions, the loss of livelihoods can lead to long-term economic stagnation, trapping entire regions in a cycle of poverty and underdevelopment.

The Intersection of Poverty and Environmental Degradation in Post-War Recovery

Environmental degradation caused by war disproportionately affects the poorest and most vulnerable populations. Those living in areas with severe contamination often lack the resources to relocate, forcing them to endure hazardous conditions that further exacerbate their poverty. The loss of access to clean water, arable land, and healthy ecosystems creates additional barriers to recovery, making it difficult for these communities to rebuild their lives.

In post-war recovery efforts, the intersection of poverty and environmental degradation poses significant challenges. Governments and aid organizations must address not only the physical damage to the environment but also the social inequalities that result from it. Poorer communities are often excluded from recovery programs, either because of a lack of resources or because they are located in less politically or economically strategic areas.

Moreover, the financial burden of environmental cleanup and healthcare often falls disproportionately on the poorest members of society. Families forced to bear the costs of medical treatment or rebuilding their homes are pushed deeper into debt, while wealthier individuals and organizations are better positioned to navigate the challenges of recovery. This inequality exacerbates existing social divides, hindering the long-term stability and development of post-war regions.

The economic and social costs of toxic legacies are among the most profound and enduring challenges of post-war recovery. From the staggering financial burden of environmental cleanup and healthcare to the displacement of communities and loss of livelihoods, the impacts of environmental degradation extend far beyond the battlefield. Addressing these issues requires a holistic approach that integrates environmental restoration, social support, and economic development, ensuring that no community is left behind in the journey toward recovery. Without such efforts, the toxic legacies of war will continue to haunt affected regions for generations.

2.5 Lessons from Past Conflicts and Their Relevance to Ukraine

The environmental devastation caused by war is not a new phenomenon, and past conflicts offer valuable lessons on how to address the toxic legacies left behind. From the chemical defoliants used in Vietnam to the oil spills of the Gulf War, history is rife with examples of how warfare reshapes landscapes and ecosystems for decades. These case studies provide insights into both the challenges of environmental restoration and the strategies that can be employed to mitigate long-term damage. For Ukraine, learning from these experiences is crucial to prevent the neglect of toxic zones and ensure a sustainable recovery.

Case Studies: Environmental Legacies of Past Wars

Vietnam War: The Devastation of Chemical Defoliants

The use of Agent Orange and other defoliants during the Vietnam War is one of the most infamous examples of environmental destruction caused by war. Millions of hectares of forests were destroyed, leaving behind barren landscapes and heavily contaminated soil. The chemicals used not only devastated ecosystems but also caused long-term health problems, including birth defects and cancers, among local populations. Despite significant international efforts to remediate affected areas, many regions remain toxic decades later.

Gulf War: Oil Spills and Fires as Ecological Catastrophes

During the Gulf War, Iraqi forces deliberately released oil into the Persian Gulf and set fire to hundreds of oil wells in Kuwait. The resulting oil spills created massive "black tides" that decimated marine life, while the smoke from the fires polluted the air and blocked sunlight for months. Cleanup efforts were monumental and costly, requiring international cooperation and advanced technologies. However, the ecological damage to marine and desert ecosystems persists, serving as a grim reminder of how deliberate environmental destruction can have lasting impacts.

Balkan Conflicts: Industrial Pollution and Urban Devastation

The wars in the Balkans during the 1990s resulted in widespread environmental damage, particularly from the targeting of industrial facilities. In Serbia, the bombing of petrochemical plants and factories released toxic chemicals into the air, soil, and water. These pollutants caused significant harm to human health and ecosystems, with many areas remaining contaminated due to insufficient cleanup efforts. The Balkans' experience highlights the long-term consequences of industrial destruction in war zones and the importance of prioritizing environmental recovery.

Approaches to Restoration: Successes and Failures from Global Examples

Vietnam's Reforestation and Cleanup Efforts

In the decades following the Vietnam War, significant efforts have been made to restore forests and clean up contaminated areas. Reforestation programs have successfully replanted large swathes of land, while international organizations have funded dioxin removal projects. However, the scale of destruction and the persistence of chemicals like Agent Orange have limited the success of these initiatives. This underscores the importance of addressing environmental damage immediately after a conflict, before it becomes entrenched.

Gulf War Cleanup: International Collaboration

The response to the environmental crises caused by the Gulf War demonstrated the value of international cooperation. Advanced technologies were used to extinguish oil fires, clean up spills, and monitor air quality. While these efforts were costly, they prevented even greater ecological damage. The Gulf War cleanup highlights the importance of mobilizing resources and expertise from multiple nations to address large-scale environmental crises.

Post-Balkan Recovery: Challenges of Limited Resources

The Balkans' experience in addressing wartime environmental damage reveals the challenges of limited resources and political instability. While some progress has been made in cleaning up industrial pollution and restoring urban areas, many toxic sites remain neglected due to a lack of funding and coordination. This serves as a cautionary tale for Ukraine, emphasizing the need for sustained investment and strong governance to ensure effective restoration.

Insights for Ukraine: Avoiding Long-Term Neglect of Toxic Zones

For Ukraine, the lessons from these past conflicts are clear: addressing environmental damage must be a central component of post-war recovery. Neglecting toxic zones not only prolongs ecological and public health crises but also undermines economic and social stability. The following insights can guide Ukraine in its efforts to restore its environment:

Prioritize Immediate Action

Delaying environmental restoration allows pollutants to spread and damage to become entrenched. Ukraine must mobilize resources to begin cleanup efforts as soon as possible, even in areas that are still experiencing sporadic conflict. Early intervention is critical to minimizing long-term costs and ensuring more effective recovery.

Leverage International Support and Expertise

Like the Gulf War cleanup, Ukraine can benefit from international cooperation. Partnering with global organizations, NGOs, and environmentally focused institutions can provide the funding, technology, and expertise needed to address large-scale environmental damage. International collaborations can also help ensure transparency and accountability in restoration efforts.

Adopt a Comprehensive Restoration Plan

A fragmented approach to addressing toxic zones will only lead to inefficiencies and missed opportunities. Ukraine needs a comprehensive, long-term plan that integrates environmental restoration with economic recovery, public health initiatives, and infrastructure rebuilding. This plan should prioritize areas with the highest levels of contamination and align with sustainable development goals.

Invest in Technology and Innovation

Advanced technologies, such as bioremediation, chemical neutralization, and remote sensing, can play a vital role in cleaning up contaminated areas. Ukraine should invest in research and development to adapt these technologies to its specific needs. Innovations in environmental restoration can also create new industries and job opportunities, contributing to the country's economic recovery.

Engage Local Communities

Restoration efforts are most successful when local communities are involved. Educating and empowering residents to participate in cleanup initiatives not only speeds up the process but also fosters a sense of ownership and resilience. Community engagement can help ensure that restoration efforts align with the needs and priorities of those most affected.

Establish Environmental Monitoring Systems

Ongoing monitoring of soil, water, and air quality is essential to track progress and identify emerging issues. Ukraine should establish robust environmental monitoring systems to ensure that cleanup efforts are effective and to prevent future contamination. These systems can also provide valuable data for international research on the long-term impacts of war on the environment.

The experiences of Vietnam, the Gulf War, and the Balkans provide a roadmap for Ukraine as it grapples with its own toxic legacies. By learning from the successes and failures of past conflicts, Ukraine has the opportunity to implement a proactive and comprehensive approach to environmental restoration. The stakes are high, but with sustained effort and global cooperation, Ukraine can transform its war-torn landscapes into foundations for a healthier, more sustainable future.

Conclusion: Toxic Legacies

The toxic legacies of war are among the most enduring and devastating consequences of modern conflict, leaving a profound imprint on the environment, human health, and socio-economic stability. The Russia-Ukraine war has underscored how chemical residues, industrial damage, and contaminated water and soil create cascading crises that extend far beyond the battlefield. These toxic remnants of war do not just poison the land; they infiltrate the lives of civilians and soldiers, disrupt ecosystems, and hinder recovery for generations.

This chapter has explored the multifaceted challenges posed by environmental degradation, from the immediate health risks of chemical exposure to the long-term economic and social costs of contaminated land

and water. The displacement of communities, the loss of livelihoods, and the chronic health impacts on vulnerable populations illustrate how deeply intertwined environmental damage is with human suffering. Furthermore, the lessons drawn from past conflicts highlight both the urgency of immediate action and the dangers of neglecting toxic zones in post-war recovery.

For Ukraine, addressing these toxic legacies is not merely an environmental imperative—it is a necessity for rebuilding resilient communities and fostering sustainable development. By learning from global examples and leveraging international cooperation, Ukraine can begin to heal its scarred landscapes and prevent these toxic legacies from becoming permanent obstacles to progress.

As the world watches the ongoing conflict, the need for comprehensive and proactive measures to address these environmental crises becomes increasingly clear. The scars of toxic legacies are not just local tragedies but global challenges, reminding us all of the true cost of war on both humanity and the planet. Only by prioritizing environmental restoration alongside other recovery efforts can Ukraine hope to transform its toxic past into a foundation for a healthier, more sustainable future.

Chapter 3: War and Climate Change

War and climate change are two of the most pressing challenges facing humanity, yet their intersection is often overlooked. Modern conflicts, such as the Russia-Ukraine war, exacerbate the climate crisis by accelerating carbon emissions, destroying vital infrastructure, and disrupting ecosystems that play a critical role in mitigating global warming. The environmental destruction caused by war not only contributes to immediate humanitarian crises but also undermines efforts to combat climate change, creating long-term challenges for the planet.

In the case of the Russia-Ukraine conflict, the targeted destruction of energy facilities, deforestation, and the emissions from military operations have significantly increased greenhouse gas levels. These impacts are not confined to the region; they ripple outward, affecting global climate stability. This chapter explores how warfare contributes to climate change, compares the Russia-Ukraine war to other global conflicts, and highlights the urgent need to address the environmental toll of war within the broader context of the climate crisis.

3.1 The Link Between War and Carbon Emissions

The connection between war and carbon emissions is an often-overlooked dimension of modern conflicts. Warfare contributes significantly to greenhouse gas emissions through the destruction of infrastructure, the energy demands of military operations, and the long-term consequences of rebuilding. In the context of global climate change, the emissions generated by conflicts like the Russia-Ukraine war exacerbate existing challenges, undermining efforts to reduce carbon footprints and stabilize the climate. This section delves into the carbon footprint of modern warfare, examines the hidden contributors to emissions, and assesses the global implications of localized conflicts.

The Carbon Footprint of Modern Warfare

Modern warfare leaves an unmistakable carbon footprint, driven by the extensive use of energy-intensive military operations and the destruction of infrastructure. From the deployment of fuel-guzzling vehicles and aircraft to the burning of oil refineries and industrial facilities, conflicts produce emissions at an alarming scale. The use of explosives, artillery, and missiles further compounds the problem, releasing greenhouse gases and particulate matter into the atmosphere.

In the Russia-Ukraine war, the destruction of critical energy infrastructure, including power plants and oil depots, has significantly increased carbon emissions. Fires from damaged refineries release massive amounts of carbon dioxide (CO_2) and black carbon into the air, while the rebuilding process requires additional energy-intensive efforts that contribute further to emissions. These immediate and long-term emissions place a significant burden on the environment, exacerbating the global climate crisis.

Weapons, Machinery, and Military Logistics: Hidden Contributors to Emissions

The tools of war—ranging from weapons and tanks to drones and aircraft—are major contributors to carbon emissions. Military machinery, often designed for power rather than efficiency, relies heavily on fossil fuels. Tanks and armored vehicles consume large quantities of diesel, while fighter jets burn vast amounts of aviation fuel during operations. Even the production and transportation of these weapons and vehicles generate significant emissions.

Logistics, a critical component of military operations, also contribute to the carbon footprint of war. The movement of troops, equipment, and supplies requires extensive use of trucks, ships, and aircraft, all of which rely on energy-intensive systems. Supply chains for fuel, food, and ammunition stretch across countries and continents, adding to the emissions associated with maintaining an active military presence.

Moreover, the production of explosives and ammunition is an energy-intensive process. Factories manufacturing these materials emit greenhouse gases, and their use in the field releases additional pollutants. For example, the detonation of bombs and missiles releases carbon dioxide, nitrogen oxides, and other harmful gases into the atmosphere, further adding to the environmental toll.

The Global Impact of Localized Conflicts on Greenhouse Gas Levels

While the emissions generated by a single conflict may seem localized, their effects are global. The interconnected nature of the Earth's atmosphere means that greenhouse gases released in one region contribute to climate change worldwide. Localized conflicts, such as the Russia-Ukraine war, act as accelerators of global warming by introducing substantial amounts of carbon dioxide and other pollutants into the atmosphere.

In addition to direct emissions, the disruption of global supply chains caused by conflicts increases energy consumption elsewhere. For example, the war in Ukraine has affected energy markets, prompting countries to turn to less sustainable energy sources, such as coal, to compensate for disrupted oil and gas supplies. This shift towards dirtier energy sources further increases global emissions, creating a cascading effect on the climate.

Furthermore, the destruction of carbon sinks—such as forests, wetlands, and grasslands—during war amplifies the impact of localized conflicts on greenhouse gas levels. These ecosystems play a vital role in absorbing carbon dioxide, and their loss reduces the planet's ability to mitigate climate change. The combined emissions from military operations, infrastructure destruction, and ecosystem loss create a global environmental crisis that transcends the boundaries of individual conflicts.

Thus, the link between war and carbon emissions highlights a critical, yet often ignored, intersection between conflict and climate change. The Russia-Ukraine war serves as a stark reminder of how military actions and their consequences contribute to global warming. Addressing this issue requires not only an understanding of the environmental toll of warfare but also a commitment to incorporating climate considerations into conflict resolution and post-war recovery efforts. Without this focus, the carbon footprint of war will continue to undermine global efforts to combat climate change, creating long-term challenges for humanity and the planet.

3.2 Destruction of Infrastructure: A Catalyst for Climate Change

Warfare frequently targets critical infrastructure, leading to catastrophic consequences for the environment and climate. The destruction of oil refineries, power plants, and other industrial facilities releases enormous amounts of greenhouse gases and pollutants into the atmosphere. These emissions not only exacerbate the immediate impacts of conflict but also create long-term challenges as nations are forced to rebuild energy-intensive facilities. The Russia-Ukraine war has highlighted the devastating impact of infrastructure destruction, emphasizing its role as a significant catalyst for climate change.

Oil Refineries and Storage Facilities: Carbon Spikes from Targeted Attacks

The targeting of oil refineries and storage facilities during conflict results in some of the most visible and severe environmental consequences. When oil refineries are bombed or set ablaze, they release vast amounts of carbon dioxide (CO_2), black carbon, and volatile organic compounds (VOCs) into the atmosphere. These pollutants not only contribute to global warming but also degrade air quality, posing serious health risks to nearby populations.

In the Russia-Ukraine war, numerous oil depots and storage tanks have been attacked, causing large-scale fires that can burn for days or weeks. The combustion of stored crude oil and refined petroleum products releases greenhouse gases at a rate far higher than standard operations. Black carbon, a short-lived climate pollutant emitted during these fires, absorbs sunlight and accelerates the melting of ice and snow when deposited on surfaces, intensifying its warming effect.

Additionally, oil spills resulting from damaged storage facilities or pipelines compound the environmental impact. Spilled oil seeps into the soil and water, disrupting ecosystems and releasing methane and other hydrocarbons as it degrades. The immediate carbon spikes caused by such attacks contribute significantly to the climate crisis, with long-lasting effects on both the local and global environment.

Power Plants in Conflict Zones: Emissions from Destruction and Disruption

Power plants, another frequent target in modern conflicts, are central to the energy infrastructure of any nation. Their destruction not only halts energy production but also releases massive amounts of carbon emissions. Coal-fired power plants, for instance, store vast amounts of ash and coal reserves on-site. When these facilities are bombed or set on fire, stored coal burns uncontrollably, emitting carbon dioxide, sulfur dioxide, and particulate matter into the atmosphere.

The disruption of power supply forces regions to rely on backup systems, such as diesel generators, which are far less efficient and produce higher emissions per unit of energy generated. In Ukraine, attacks on power plants have led to widespread blackouts, prompting increased reliance on emergency energy solutions that contribute to localized air pollution and higher carbon footprints.

Nuclear power plants, while not directly emitting greenhouse gases during operations, pose unique risks during conflict. Damage to these facilities can lead to radioactive contamination, necessitating the shutdown of operations and forcing countries to rely on fossil fuels to compensate for the energy shortfall. This shift to carbon-intensive energy sources has a profound impact on emissions, exacerbating climate change in the absence of cleaner energy alternatives.

Long-Term Climate Implications of Rebuilding Damaged Infrastructure

The destruction of critical infrastructure during war creates a long-term environmental burden as nations undertake the immense task of rebuilding. The reconstruction of oil refineries, power plants, and industrial facilities is an energy-intensive process that requires large quantities of materials such as steel, cement, and glass—industries that are themselves significant contributors to greenhouse gas emissions.

In the aftermath of conflict, nations often prioritize speed and cost over sustainability, leading to the rapid construction of facilities that rely on outdated, carbon-intensive technologies. This approach locks regions into high-emission energy systems for decades, undermining global efforts to transition to renewable energy and mitigate climate change.

Moreover, the loss of pre-existing renewable energy infrastructure, such as wind farms or solar installations, further exacerbates the problem. Replacing these facilities can take years, during which nations often fall back on coal and oil to meet their energy demands, leading to a spike in emissions. The destruction of infrastructure in Ukraine, for example, has delayed the country's progress toward its climate goals, forcing a reliance on emergency solutions that increase its carbon footprint.

Another long-term implication is the financial burden of rebuilding, which often diverts resources away from investments in renewable energy and climate adaptation measures. Post-war economies, struggling to recover, are less likely to prioritize sustainable energy solutions, perpetuating cycles of carbon-intensive growth.

In summary, the destruction of infrastructure during war is a critical driver of climate change, with immediate emissions from fires and explosions compounded by the long-term environmental costs of rebuilding. The Russia-Ukraine war has demonstrated the devastating climate impact of targeted attacks on oil refineries, power plants, and other facilities, emphasizing the urgent need to address these issues in post-war recovery efforts. By incorporating sustainability into rebuilding processes and transitioning to cleaner energy systems, nations can mitigate the climate consequences of conflict and contribute to a more resilient future. Without such measures, the environmental toll of infrastructure destruction will continue to fuel the global climate crisis.

3.3 War-Induced Deforestation and Loss of Carbon Sinks

Forests are among the most significant casualties of war, often destroyed as collateral damage during military operations. These natural ecosystems play a critical role as carbon sinks, absorbing vast amounts of carbon dioxide (CO_2) and mitigating global warming. However, during conflicts, deforestation and land clearing not only eliminate this vital function but also release large amounts of stored carbon into the atmosphere, accelerating climate change. The Russia-Ukraine war, like many other conflicts, has highlighted the devastating consequences of war-induced deforestation, with implications that extend far beyond the battlefield.

Forests as Collateral Damage: The Dual Impact of Deforestation

Forests frequently become unintended casualties during war, falling victim to bombings, artillery fire, and military encampments. In many cases, they are deliberately cleared for tactical purposes, such as removing cover for enemy forces or creating open fields of visibility for combat. These actions result in the destruction of vegetation, soil degradation, and the disruption of entire ecosystems.

The dual impact of deforestation during war is twofold. First, the loss of trees eliminates their ability to act as carbon sinks, depriving the planet of one of its most effective tools for capturing atmospheric CO_2. Forests are essential for maintaining the Earth's carbon balance, and their destruction significantly reduces global carbon sequestration capacity.

Second, the burning or clearing of forests releases the carbon stored within trees and soil. This carbon, which has often been sequestered for decades or even centuries, is rapidly released into the atmosphere in the form of CO_2, contributing to global warming. The immediate emissions from deforestation during war exacerbate the already significant carbon footprint of military operations, making conflicts a major driver of climate change.

The Release of Stored Carbon from Burned and Cleared Land

The clearing and burning of forests during wartime are particularly damaging because of the massive release of stored carbon. When forests are destroyed, not only do the trees release their stored carbon, but the soil itself becomes a source of emissions. Soil holds vast amounts of organic carbon, which is released into the atmosphere when it is disturbed or exposed.

Forest fires caused by military activity, such as shelling or airstrikes, amplify this problem. The intense heat from these fires releases carbon dioxide, methane, and other greenhouse gases into the atmosphere. Methane, in particular, has a much greater warming potential than CO_2, making these emissions even more harmful in the short term.

For example, during the Russia-Ukraine war, military activity in forested regions has resulted in large-scale fires, particularly in areas where battles are intense or where fuel depots and other flammable materials are stored near forests. These fires not only destroy the forests themselves but also disrupt nearby peatlands and wetlands, which are significant carbon reservoirs. The release of carbon from these ecosystems contributes to a feedback loop of warming and environmental degradation.

Case Studies: Forest Loss During the Russia-Ukraine War and Other Conflicts

Russia-Ukraine War: Targeting Forested Regions

The Carpathian Mountains and forested areas near the Donbas region in Ukraine have been heavily affected by the ongoing conflict. These forests, which act as crucial carbon sinks for the region, have been damaged by shelling, military encampments, and accidental fires. The loss of forest cover in these areas has disrupted local ecosystems and released substantial amounts of carbon dioxide into the atmosphere.

Additionally, the destruction of shelterbelt forests, which were planted to protect farmland from erosion and to maintain soil quality, has reduced agricultural productivity and further increased emissions. The cumulative impact of forest loss during this war contributes not only to Ukraine's environmental challenges but also to global carbon emissions.

Vietnam War: Chemical Defoliation and Ecosystem Collapse

During the Vietnam War, the use of Agent Orange and other defoliants led to the destruction of millions of hectares of forest. This deliberate targeting of vegetation to deny cover to enemy forces resulted in the release of massive amounts of stored carbon, alongside the long-term degradation of soil and ecosystems. The environmental consequences of this deforestation continue to affect Vietnam today, with many areas unable to recover their original biodiversity or carbon sequestration capacity.

Syrian Civil War: Deforestation for Fuel and Survival

In Syria, prolonged conflict has forced local populations to turn to forests for survival, cutting down trees for fuel and shelter. This wartime deforestation, driven by necessity, has reduced forest cover in many parts of the country, diminishing their ability to act as carbon sinks. The cumulative effect of such practices, coupled with damage caused by military activity, has contributed to environmental degradation and increased emissions in the region.

Gulf War: Burning of Forests and Desertification

Although primarily known for the burning of oil fields, the Gulf War also saw the destruction of forested areas and vegetation due to military movements and bombing campaigns. The loss of these natural carbon sinks, combined with the release of greenhouse gases from fires, had significant implications for the region's carbon balance.

The Broader Implications of War-Induced Deforestation

The destruction of forests during conflicts like the Russia-Ukraine war has far-reaching implications for both local environments and the global climate. Forests are irreplaceable in their role as carbon sinks, biodiversity hotspots, and protectors of soil and water systems. Their loss accelerates climate change, disrupts ecosystems, and undermines the resilience of affected regions to recover from environmental and economic shocks.

Reforestation and afforestation efforts are critical to addressing these challenges, but they require time, resources, and political will. Lessons from past conflicts demonstrate the importance of integrating forest restoration into post-war recovery plans and prioritizing the protection of remaining forests during wartime. Without such measures, the environmental and climate impacts of war-induced deforestation will continue to compound, leaving a lasting legacy of destruction for future generations.

3.4 Global Comparisons: Lessons from Other Conflicts

The environmental impacts of the Russia-Ukraine war are not isolated incidents. Previous conflicts around the world have left behind significant ecological damage, providing valuable lessons on how war contributes to climate change and environmental degradation. From the burning oil fields of the Gulf War to the chemical defoliation during the Vietnam War and resource depletion in Syria, each conflict highlights distinct yet interconnected pathways through which warfare exacerbates environmental crises. By examining these cases, we can better understand the broader implications of war on the environment and draw parallels to ongoing challenges in Ukraine.

The Gulf War: Burning Oil Fields and Their Climate Legacy

The Gulf War of 1990-1991 remains one of the most environmentally catastrophic conflicts in modern history. During their retreat, Iraqi forces set fire to over 700 oil wells in Kuwait, creating massive infernos that burned for months. The fires released enormous amounts of carbon dioxide (CO_2), black carbon, and other greenhouse gases into the atmosphere, significantly contributing to global warming. It is estimated that the burning oil wells released as much as 500 million tons of CO_2, equivalent to the annual emissions of many industrialized nations.

The smoke from the fires darkened the skies across the Gulf region, reducing sunlight penetration and disrupting local weather patterns. Black carbon particles from the fires were carried by wind to distant regions, where they contributed to the melting of glaciers and snow in higher altitudes. The release of toxic gases such as sulfur dioxide also led to acid rain, further damaging ecosystems and agricultural lands.

Efforts to extinguish the fires required international cooperation, but the environmental consequences persisted for years. The region's soil and water sources were heavily contaminated by oil spills and residues, rendering large tracts of land unproductive. The Gulf War serves as a stark reminder of how targeted attacks on energy infrastructure can have immediate and long-term climate repercussions.

Vietnam: The Environmental Toll of Chemical Defoliants and Carbon Release

The Vietnam War (1955-1975) saw one of the most deliberate and large-scale uses of environmental destruction as a military strategy. The U.S.

military deployed chemical defoliants such as Agent Orange to strip forests of their foliage, depriving enemy forces of cover and disrupting food supplies. Over 19 million gallons of herbicides were sprayed across Vietnam, affecting nearly five million hectares of forest and farmland.

This deforestation not only devastated ecosystems but also released vast amounts of stored carbon into the atmosphere. Forests, which act as natural carbon sinks, were destroyed at an unprecedented scale, contributing to increased greenhouse gas levels. The long-term impact on Vietnam's soil and water quality has further hampered reforestation efforts, prolonging the carbon deficit created during the war.

The use of chemical defoliants caused irreversible damage to biodiversity, with many plant and animal species disappearing from affected regions. The ecological damage also had profound social and economic consequences, as rural communities dependent on forest resources faced long-term displacement and loss of livelihoods.

Unlike the Gulf War, where efforts were made to address environmental damage post-conflict, the environmental toll of the Vietnam War has received limited remediation. While some reforestation projects have been initiated, the scars left by chemical defoliation remain visible to this day, demonstrating the lasting climate and ecological consequences of such warfare strategies.

Syria: Conflict-Induced Resource Depletion and Its Climate Implications

The ongoing Syrian Civil War, which began in 2011, illustrates a different but equally damaging form of environmental impact: resource depletion. Prolonged conflict in Syria has led to widespread overexploitation of natural resources, including forests, water, and agricultural land, as communities struggle to survive amidst war. The deforestation of Syria's remaining woodlands has accelerated during the conflict, as displaced populations rely on wood for fuel and construction. This loss of forest cover has reduced the country's capacity to absorb carbon dioxide, amplifying its contribution to climate change.

Water resources in Syria have also been severely affected. The destruction of irrigation systems, contamination of water supplies, and overuse of groundwater have created a water crisis that not only impacts human survival but also disrupts agricultural production. This, in turn, has led to food insecurity and forced migration, further stressing regional ecosystems and increasing carbon emissions associated with displacement.

The war has also disrupted Syria's already fragile energy infrastructure, leading to a reliance on inefficient and carbon-intensive methods of energy production, such as diesel generators. The economic challenges of rebuilding and restoring sustainable energy systems in a post-conflict scenario add another layer to the environmental toll of the war.

Unlike the Gulf War or Vietnam, where the primary environmental damage was linked to direct military actions, Syria highlights the compounded effects of prolonged conflict on natural resources and ecosystems. The degradation of these resources exacerbates climate vulnerabilities, creating a feedback loop of environmental and social challenges.

Relevance to the Russia-Ukraine War

The lessons from these global conflicts provide valuable insights for understanding the environmental impact of the Russia-Ukraine war. Like the Gulf War, targeted attacks on energy infrastructure in Ukraine have led to significant greenhouse gas emissions. The destruction of forests and agricultural land in Ukraine parallels the ecological devastation seen in Vietnam, emphasizing the long-term carbon consequences of deforestation. Meanwhile, the depletion of resources in conflict zones like Syria serves as a warning of how prolonged war can destabilize ecosystems and amplify climate challenges.

These case studies also highlight the importance of proactive environmental recovery efforts. While the Gulf War demonstrated the potential for international cooperation in addressing environmental crises, the limited remediation in Vietnam and Syria underscores the consequences of neglect. For Ukraine, addressing the environmental toll of the war will require immediate action, sustained investment, and global support to mitigate the long-term impacts on its ecosystems and climate resilience.

3.5 The Ripple Effects of War on Climate Policies

War and conflict have a profound and often overlooked impact on global climate policies. The resources, attention, and political will that are diverted to resolving conflicts come at the expense of advancing environmental and climate action. Moreover, wars disrupt the momentum of renewable energy transitions, stall environmental goals, and exacerbate emissions, complicating international efforts to combat climate change. The Russia-Ukraine war serves as a striking example of how armed conflict undermines climate policies and challenges the global response to the climate crisis.

The Diversion of Resources from Climate Action to Conflict Resolution

One of the most immediate ripple effects of war on climate policies is the diversion of financial and human resources. Governments and international organizations that might otherwise invest in renewable energy projects, sustainable infrastructure, or climate adaptation initiatives are forced to redirect funds and efforts to conflict resolution and humanitarian aid.

For example, the billions of dollars spent on arms, military aid, and rebuilding war-torn regions represent resources that could have been allocated to achieving global climate goals. In Ukraine, the costs of addressing infrastructure destruction and humanitarian crises have overshadowed investments in green energy and climate resilience. Similarly, countries providing aid to Ukraine face domestic pressures to prioritize defense and geopolitical interests over environmental spending.

This diversion also affects developing nations, which are already struggling to finance climate adaptation and mitigation. With global financial resources tied up in conflict-related expenditures, fewer funds are available for vulnerable countries to address rising sea levels, extreme weather events, and other climate-related challenges. The result is a domino effect that undermines progress on climate action at both local and global levels.

War as a Setback for Renewable Energy Adoption and Environmental Goals

Conflicts disrupt the adoption of renewable energy technologies and hinder progress toward environmental goals. The destruction of energy infrastructure during war often forces countries to fall back on fossil fuels as a temporary solution, increasing carbon emissions. For instance, Ukraine's reliance on coal and gas-fired power plants has grown due to the destruction of renewable energy facilities and the instability of energy supplies during the conflict.

Wars also create logistical and economic barriers to renewable energy development. Supply chains for solar panels, wind turbines, and battery technologies are disrupted, delaying the deployment of renewable energy projects. In addition, economic instability caused by conflict discourages investment in long-term green energy solutions, as governments and private sectors prioritize immediate recovery over sustainable development.

The Russia-Ukraine war has also had a significant impact on global energy markets, leading to increased reliance on carbon-intensive fuels. European nations, seeking to reduce their dependence on Russian gas, have turned to coal and liquefied natural gas (LNG) as stopgap measures, undermining their commitments to phase out fossil fuels. This shift has delayed the transition to clean energy, jeopardizing the EU's ambitious climate targets.

Global Climate Agreements and Their Response to War-Induced Emissions

International climate agreements, such as the Paris Agreement, aim to unite nations in addressing climate change. However, the emergence of large-scale conflicts complicates the implementation of these agreements. War-induced emissions are often not accounted for in national carbon inventories, creating gaps in climate accountability. The emissions resulting from the destruction of infrastructure, increased military activity, and emergency energy production go largely unreported, making it difficult to assess the true impact of conflicts on global warming.

The Russia-Ukraine war has posed a significant challenge to global climate diplomacy. While many nations remain committed to their climate goals, the geopolitical tensions and economic disruptions caused by the war have stalled international cooperation. Global summits like COP have struggled to address the environmental consequences of war, as they are typically focused on peacetime strategies for reducing emissions.

Moreover, the war has highlighted the limitations of current climate frameworks in dealing with conflict-related emissions. There is a growing need for mechanisms within global agreements to account for and mitigate the environmental impacts of war. This includes strategies for rebuilding green infrastructure in post-conflict regions and ensuring that emissions from military activities are included in climate reporting.

Relevance to the Russia-Ukraine War

The Russia-Ukraine war illustrates how conflict can derail progress on climate action and undermine global agreements. The diversion of resources, the resurgence of fossil fuel reliance, and the lack of accountability for war-related emissions highlight critical gaps in climate policies. For Ukraine, rebuilding its energy systems offers an opportunity to prioritize renewable energy, but this will require substantial international support and funding.

On a global scale, the war serves as a wake-up call for integrating conflict considerations into climate policies. This includes developing frameworks to address war-induced emissions, ensuring that post-war reconstruction aligns with environmental goals, and safeguarding progress on renewable energy transitions during periods of instability. Without these measures, conflicts will continue to erode the collective efforts needed to combat climate change, leaving the planet more vulnerable to its consequences.

Thus, the ripple effects of war on climate policies underscore the interconnectedness of global challenges. As conflicts like the Russia-Ukraine war disrupt energy systems, divert resources, and stall international cooperation, the need for resilient and adaptive climate policies becomes increasingly urgent. Addressing these impacts requires a comprehensive approach that prioritizes sustainability even in the face of conflict, ensuring that climate action remains a global priority in an increasingly turbulent world.

Interview with Dr. Lucas Fernandis, Climate Policy Expert

Dr. Lucas Fernandis is a renowned climate policy expert with extensive experience advising international organizations and governments on sustainable development and energy transition strategies. His work focuses on the intersection of climate policy and global security, addressing how geopolitical conflicts impact climate goals.

Q: Dr. Fernandis, thank you for taking the time to speak with us. Let's start with the broader implications of war, like the Russia-Ukraine conflict, on global climate efforts. How does war disrupt progress toward climate goals?

A: Thank you for having me. War fundamentally disrupts climate progress in several ways. First, conflicts divert attention and resources from long-term climate goals to immediate crises. Governments shift funding away from renewable energy projects and sustainability initiatives toward defense budgets and emergency relief efforts. This redirection slows the momentum of global climate action.

Second, wars often lead to an increased reliance on fossil fuels. For example, the Russia-Ukraine conflict has caused significant disruptions to global energy markets, forcing countries to reopen coal plants or import oil and gas to compensate for lost supplies. This backpedaling undermines efforts to reduce carbon emissions and transition to cleaner energy sources.

Lastly, conflicts create environmental destruction that exacerbates climate challenges. Damaged infrastructure, deforestation, and pollution release vast amounts of greenhouse gases into the atmosphere. In the case of Ukraine, attacks on industrial sites and energy grids have resulted in widespread emissions, adding to the global carbon footprint.

Q: How has the Russia-Ukraine conflict, specifically, influenced international climate policies?

A: The Russia-Ukraine conflict has had both direct and indirect effects on international climate policies. Directly, the war has shifted the focus of many European nations from ambitious climate commitments to securing immediate energy security. Countries that had pledged to reduce their reliance on fossil fuels have had to increase coal usage and ramp up LNG imports to address shortages caused by reduced Russian gas supplies.

Indirectly, the conflict has underscored the vulnerability of global energy systems. It's a wake-up call for many nations about the importance of diversifying their energy sources and investing in renewables. We've seen some positive responses, such as accelerated wind and solar projects in Europe, but these efforts are often overshadowed by short-term emergency measures.

At the international level, the war has made global climate negotiations more challenging. Divisions between nations have deepened, and the focus on geopolitical rivalries has reduced the urgency of collective action on climate change. For instance, discussions during COP summits have been sidetracked by geopolitical tensions, diluting the momentum for meaningful agreements.

Q: Could the push for energy security in response to conflicts like this inadvertently support renewable energy adoption?

A: Absolutely, and this is one of the silver linings amidst the challenges. Energy security and climate goals don't have to be at odds—they can be mutually reinforcing. The disruption of fossil fuel supplies has highlighted the risks of over-reliance on centralized, non-renewable energy systems. Many nations are now fast-tracking renewable energy projects to reduce their dependency on volatile fossil fuel markets.

For example, Germany, which faced significant energy shortages during the conflict, has expedited its offshore wind and solar energy projects. Similarly, the European Union has proposed ambitious plans to expand renewable energy generation as part of its REPowerEU initiative, which aims to reduce reliance on Russian energy.

However, the key is ensuring that these efforts are sustained and not sidelined once the immediate crisis subsides. The transition to renewables must remain a priority even as energy markets stabilize.

Q: What role can international organizations play in maintaining focus on climate goals during geopolitical conflicts?

A: International organizations have a critical role to play in ensuring that climate action remains on the global agenda, even during conflicts. They can:

1. **Promote Collaboration Across Borders:** Organizations like the United Nations and the International Renewable Energy Agency (IRENA) can facilitate cross-border renewable energy projects that enhance regional energy security while reducing emissions.
2. **Provide Funding for Resilience:** Institutions like the World Bank and the Green Climate Fund should prioritize funding for nations affected by conflicts to rebuild sustainably. This includes investments in renewable energy infrastructure and climate adaptation projects.
3. **Set Clear Benchmarks:** International organizations must hold countries accountable for their climate commitments, ensuring that short-term measures taken during conflicts don't derail long-term goals.

These organizations can also act as mediators, helping nations navigate the delicate balance between energy security and sustainability during crises.

Q: What lessons can be learned from the Russia-Ukraine conflict to better align geopolitical and climate priorities?

A: One of the key lessons is the importance of **resilience**. The conflict has shown us that energy systems need to be resilient not just to climate risks but also to geopolitical disruptions. Investments in decentralized energy systems, such as rooftop solar and community wind projects, can reduce the vulnerability of national grids to conflict-related disruptions.

Another lesson is the value of **diversification**. Relying heavily on a single energy source or supplier creates systemic vulnerabilities. Nations must diversify their energy mix, incorporating renewables alongside robust storage and distribution systems.

Finally, the conflict highlights the need for **global cooperation**. Climate change and geopolitical conflicts are interconnected challenges that require collective solutions. Whether it's through technology transfer, joint investments in renewable projects, or shared energy grids, international collaboration is essential to align geopolitical and climate priorities.

Q: What message would you share with policymakers balancing immediate energy security concerns with long-term climate goals?

A: My message is this: Short-term energy security measures must not come at the expense of long-term climate goals. It's understandable that during a crisis, governments focus on stabilizing energy supplies, but they must simultaneously lay the groundwork for a sustainable future. Every investment made in fossil fuels today should be matched or exceeded by investments in renewables, storage, and grid modernization.

Policymakers should also recognize that addressing climate change is a matter of national and global security. The impacts of climate instability—food shortages, resource conflicts, mass migrations—are just as destabilizing as traditional geopolitical threats. By prioritizing sustainability, nations can build a foundation for both energy security and climate resilience.

Dr. Lucas Fernandis' insights highlight the intricate relationship between geopolitical conflicts and climate action. His expertise underscores the urgency of aligning energy security with sustainability, ensuring that the crises of today do not undermine the progress needed for a stable and resilient future.

3.6 Ukraine's Role in the Climate Equation

Ukraine holds a unique position in the global climate equation due to its rich natural resources, agricultural significance, and potential for renewable energy development. However, the ongoing conflict has severely disrupted its ability to contribute to global climate initiatives, turning its environmental assets into collateral damage. Balancing the demands of post-war recovery with the need for sustainable development presents a long-term challenge, not just for Ukraine but for the broader global community. This section explores Ukraine's environmental significance, the impact of the war on its climate contributions, and the critical steps needed to align recovery efforts with sustainability.

The Environmental Significance of Ukraine's Natural Resources

Ukraine's natural resources are a cornerstone of its environmental and economic landscape. Known as Europe's "breadbasket," Ukraine possesses some of the most fertile agricultural lands in the world, producing a significant portion of the globe's wheat, corn, and sunflower oil. These agricultural systems are not only vital for global food security but also for mitigating climate change, as healthy soils act as carbon sinks, sequestering carbon dioxide from the atmosphere.

In addition to its agricultural wealth, Ukraine is home to vast forests, wetlands, and biodiversity-rich ecosystems that play a crucial role in regulating regional and global climates. Forests in regions like the Carpathians act as natural buffers against climate change, absorbing CO_2 while supporting a diverse array of plant and animal life.

Ukraine also has significant untapped potential for renewable energy. Its geographic location and natural conditions make it well-suited for wind, solar, and bioenergy projects, positioning the country as a potential leader in clean energy development. Prior to the war, Ukraine had begun transitioning its energy systems, with ambitious goals to reduce greenhouse gas emissions and integrate renewables into its energy mix.

How War Disrupts Ukraine's Contribution to Global Climate Initiatives

The ongoing conflict has dramatically undermined Ukraine's ability to contribute to global climate initiatives. The war has resulted in the large-scale destruction of natural ecosystems, agricultural lands, and renewable energy infrastructure, reversing years of progress in environmental sustainability.

Ecosystem Destruction: Bombings, deforestation, and pollution have severely impacted Ukraine's natural ecosystems. Forest fires caused by military activity have destroyed carbon-absorbing vegetation, while chemical spills and contamination have degraded wetlands and waterways. The loss of these ecosystems diminishes their ability to act as carbon sinks, increasing greenhouse gas levels and disrupting biodiversity.

Agricultural Disruption: The war has displaced farmers, destroyed farmlands, and disrupted supply chains, reducing Ukraine's agricultural output. This not only impacts global food security but also hampers sustainable agricultural practices. Soil degradation and contamination caused by the conflict have further reduced the land's capacity to sequester carbon, making the recovery of agricultural systems an urgent climate priority.

Energy Infrastructure Damage: Attacks on power plants, renewable energy facilities, and energy storage systems have forced Ukraine to rely more heavily on fossil fuels to meet immediate energy demands. This shift undermines the country's progress toward decarbonization, increasing its carbon footprint at a time when global climate commitments require rapid reductions in emissions.

Disruption to International Commitments: As a signatory to the Paris Agreement, Ukraine had committed to ambitious climate goals, including reducing greenhouse gas emissions by 65% by 2030 compared to 1990 levels. The war has derailed these efforts, making it nearly impossible for Ukraine to meet its targets. Additionally, the focus on conflict resolution has diverted attention and resources away from climate initiatives, delaying critical progress.

The Long-Term Challenge of Balancing Recovery with Sustainability

Rebuilding Ukraine in the aftermath of war presents a significant opportunity to align recovery efforts with sustainability. However, achieving this balance will require careful planning, substantial investment, and international cooperation.

Sustainable Reconstruction: The reconstruction of Ukraine's infrastructure must prioritize energy efficiency and renewable energy integration. Rebuilding destroyed power plants and industrial facilities offers an opportunity to adopt green technologies, reducing the country's reliance on fossil fuels and aligning with global decarbonization goals. Investing in wind and solar energy projects, as well as modernizing energy grids, will be critical to this effort.

Restoring Ecosystems: The recovery of Ukraine's natural ecosystems is essential for both environmental and economic stability. Large-scale reforestation projects, wetland restoration, and soil rehabilitation can help restore the country's carbon sequestration capacity while protecting biodiversity. These efforts must be supported by comprehensive monitoring systems to ensure the health and resilience of restored ecosystems.

Revitalizing Agriculture: Sustainable agricultural practices must be at the forefront of Ukraine's recovery. Encouraging the adoption of regenerative farming techniques, such as crop rotation, reduced tillage, and organic farming, can help rebuild soil fertility and reduce emissions from the agricultural sector. International partnerships and investments will be necessary to provide farmers with the tools and knowledge needed to transition to sustainable practices.

Global Support and Integration: Ukraine's recovery will require substantial international support, both in terms of funding and technical expertise. Global climate frameworks, such as the Green Climate Fund, can play a pivotal role in financing sustainable recovery projects. Additionally, integrating Ukraine into international clean energy markets and climate initiatives can help accelerate its transition to a green economy.

Addressing Social and Economic Challenges: Sustainable recovery efforts must also address the social and economic impacts of the war. Ensuring that local communities benefit from green reconstruction projects, creating jobs in renewable energy sectors, and promoting environmental education are key steps in building a resilient and inclusive recovery.

Relevance to the Global Climate Equation

Ukraine's role in the climate equation extends beyond its borders. The country's natural resources, agricultural systems, and renewable energy potential have global implications for food security, biodiversity, and greenhouse gas mitigation. The war has not only disrupted these contributions but also highlighted the need for integrating climate resilience into post-conflict recovery efforts.

As the world grapples with the dual challenges of conflict and climate change, Ukraine's experience underscores the importance of aligning recovery efforts with sustainability. By rebuilding green, Ukraine can serve as a model for integrating environmental goals into post-war reconstruction, contributing to global climate stability and setting a precedent for other nations facing similar challenges.

The stakes are high, but with the right investments and international collaboration, Ukraine can transform its recovery into an opportunity for sustainable growth, reinforcing its role as a key player in the global fight against climate change.

Conclusion: War and Climate Change

The intersection of war and climate change is a sobering reminder of how modern conflicts exacerbate environmental crises. The Russia-Ukraine war has revealed the profound impacts that warfare can have on carbon emissions, renewable energy progress, and global climate policies. From the destruction of critical infrastructure to the loss of carbon sinks through deforestation, the conflict underscores the complex relationship between war and the accelerating climate crisis.

This chapter has explored how conflicts disrupt nations' contributions to global climate initiatives, divert resources from environmental priorities, and deepen reliance on fossil fuels. The ripple effects of war-induced emissions extend beyond national borders, affecting global efforts to combat climate change. Ukraine's critical role in the climate equation, as both a contributor to food security and a potential leader in renewable energy, emphasizes the urgent need for sustainable recovery.

The path forward must integrate climate resilience into post-war rebuilding efforts. International collaboration, investment in green technologies, and restoration of ecosystems will be essential to mitigating the long-term climate impacts of the war. The lessons drawn from global conflicts also highlight the necessity of addressing war's environmental toll within international climate agreements and frameworks.

Ultimately, the Russia-Ukraine war serves as a stark reminder of how intertwined the challenges of conflict and climate change are. Addressing these dual crises requires not only immediate action but also a commitment to rebuilding in a way that prioritizes sustainability and global climate goals. By doing so, Ukraine has the opportunity to emerge as a model for green recovery, contributing to a more resilient and sustainable future for the planet.

Interview with Oleksandr Hrytsenko, Ukrainian Farmer and War Survivor

As part of understanding the environmental and economic challenges faced by Ukraine due to the ongoing conflict, we spoke to **Oleksandr Hrytsenko**, a farmer from the Chernihiv region. His farm, once a thriving agricultural hub, is now a grim reflection of how war has upended livelihoods and disrupted Ukraine's role in the global climate equation.

Interviewer: Oleksandr, thank you for taking the time to speak with us. Could you start by describing your life before the war?

Oleksandr: Before the war, my family and I managed about 50 hectares of farmland in Chernihiv. We grew wheat, sunflowers, and some vegetables. This land has been in my family for generations, and farming wasn't just a livelihood—it was our way of life. We had invested in sustainable farming techniques, using crop rotation and minimal chemical inputs to keep the soil healthy. I was proud of what we were contributing to Ukraine's reputation as the "breadbasket of Europe."

Interviewer: How has the war affected your farm and your ability to continue farming?

Oleksandr: The impact has been devastating. When the fighting reached our region, my farm became part of the battleground. Shelling destroyed the irrigation systems and storage facilities we had built over decades. Some of my fields were littered with unexploded ordnance, making them too dangerous to use. On top of that, chemicals from nearby industrial sites targeted during airstrikes leaked into the soil and nearby water sources. Even now, I don't know if the land will ever be safe for farming again.

Interviewer: That sounds incredibly challenging. What has been the impact on the local environment and agriculture in your area?

Oleksandr: The soil is poisoned, the rivers are polluted, and much of the farmland has been abandoned. When I walk through what used to be lush fields, I see craters where bombs have fallen and burnt patches where fires destroyed crops. Wildlife has disappeared—no birds, no small animals. It's eerie. For farmers, the loss isn't just financial; it's emotional. This land was our connection to nature and to our heritage.

Interviewer: Ukraine has played a significant role in global food production and climate initiatives. How do you see the war affecting these contributions?

Oleksandr: Before the war, Ukraine was feeding millions worldwide and working toward more sustainable farming practices. Now, much of our agricultural output has been disrupted, and the idea of sustainability feels like a distant dream. With damaged fields and contaminated soil, we're forced to use whatever methods we can just to produce something. Unfortunately, this means relying on more chemicals, which isn't good for the environment or the long-term health of the land.

But it's not just about food production. Ukraine's farmland is a major carbon sink. If we can't restore it, we'll be contributing to global emissions instead of reducing them. The world needs to see this not just as a Ukrainian problem but as a global one.

Interviewer: What kind of support do you think farmers like yourself need to recover and contribute to a sustainable future?

Oleksandr: First and foremost, we need the land to be made safe. Clearing unexploded ordnance and decontaminating the soil has to be the top priority. Without that, we can't rebuild. After that, we'll need financial support and access to resources to restart farming. But more than just farming, we need to think about sustainability.

If international organizations can help us rebuild irrigation systems, provide clean energy solutions for our farms, and offer training in regenerative agriculture, we can come back stronger. The war has taken so much from us, but it's also shown us how fragile our systems are. If we can rebuild in a way that respects the land and the climate, we can turn this tragedy into an opportunity.

Interviewer: That's a powerful vision. Lastly, do you have a message for the international community?

Oleksandr: Yes. Don't forget about Ukraine. We're not just fighting for our survival; we're also fighting for a future where our land can feed people and help combat climate change. The world's support isn't just charity—it's an investment in a more sustainable future for everyone. I hope the international community sees what's happening here and works with us to rebuild not just our farms but our role in protecting the planet.

This interview sheds light on the personal struggles of Ukrainian farmers like Oleksandr, who are on the frontlines of both the war and its environmental consequences. His story underscores the importance of integrating sustainability into Ukraine's recovery and ensuring that the world recognizes the broader implications of the conflict on climate and global food security.

Chapter 4: The Forgotten Victims: Wildlife and Habitats

Amid the human suffering and infrastructural destruction caused by war, the plight of wildlife and natural habitats often goes unnoticed. Yet, animals and ecosystems are among the most vulnerable victims of conflict. The chaos of war disrupts ecosystems, destroys protected areas, and forces the migration or death of countless species. Endangered animals lose their habitats, while conservation efforts are brought to a halt, leaving these "forgotten victims" to fend for themselves in increasingly hostile environments.

In the Russia-Ukraine war, forests, wetlands, and biodiversity hotspots have become unintended battlegrounds, threatening unique species and fragile ecosystems. Animals are displaced by explosions, poisoned by pollutants, and trapped by fragmented landscapes, leading to ripple effects that destabilize entire ecosystems. This chapter explores the devastating impact of war on wildlife and habitats, the challenges of conservation during conflict, and the long-term implications for biodiversity and ecological recovery.

4.1 Impact of War on Endangered Species and Protected Areas

Wars are not confined to cities and human settlements; they often spill over into forests, wetlands, and other natural habitats, devastating ecosystems and threatening the survival of countless species. Endangered animals and protected areas, which rely on stability and conservation efforts, become unintended victims of conflict. In the Russia-Ukraine war, as in other conflicts, biodiversity hotspots and nature reserves have suffered significant destruction, putting unique species and fragile ecosystems at risk.

Destruction of Biodiversity Hotspots and Nature Reserves

Protected areas, which are specifically designated to safeguard ecosystems and species, are particularly vulnerable during war. Military operations destroy habitats through bombings, shelling, and troop movements, leaving once-thriving ecosystems scarred and fragmented. Forests are cleared to make way for military operations, wetlands are drained or polluted, and grasslands are turned into battlegrounds, reducing the availability of critical habitats for species.

In Ukraine, regions like the Polissya Biosphere Reserve and the Carpathian Biosphere Reserve—home to diverse flora and fauna—have been severely impacted by the conflict. Fires ignited by shelling have destroyed large sections of forests, while pollutants from bombings and industrial damage have seeped into rivers and wetlands, further degrading these ecosystems. Such destruction not only affects local wildlife but also disrupts the ecological balance of larger regions, as protected areas often serve as vital corridors for migration and genetic exchange.

Threats to Endangered Species from Habitat Loss and Poaching

For endangered species, the destruction of habitats caused by war can mean the difference between survival and extinction. Many of these species are already teetering on the edge due to habitat fragmentation, climate change, and human activity. The chaos of war exacerbates these pressures, stripping them of the food, shelter, and stability they need to survive.

In conflict zones, the enforcement of anti-poaching laws becomes nearly impossible, leaving endangered species vulnerable to exploitation. Poaching often increases during wars, as people turn to hunting for survival or exploit the absence of law enforcement to harvest valuable animal products, such as ivory, pelts, or exotic pets. In Ukraine, the Eurasian lynx, European bison, and black stork, which are already under threat, face additional challenges as their habitats are destroyed and their movements disrupted by military activities.

Even species that are not traditionally targeted by poachers suffer during conflicts. Destruction of nesting sites, contamination of food sources, and disruption of breeding patterns can have long-term consequences for population recovery. For example, migratory birds that rely on wetlands in Ukraine face not only the loss of their habitats but also heightened risks during their seasonal journeys, as war zones create barriers to their natural routes.

Case Studies: Wildlife Casualties in Ukraine's Protected Areas

The Russia-Ukraine war has left visible scars on some of the country's most ecologically significant areas, with direct consequences for wildlife. Below are key examples of how the conflict has impacted protected areas and their inhabitants:

The Polissya Biosphere Reserve: Spanning Ukraine and Belarus, the Polissya Biosphere Reserve is a critical habitat for rare species such as the European bison and greater spotted eagle. Military activity in this area has caused extensive damage to forests and wetlands. Explosions and fires have destroyed vegetation, while pollutants have contaminated the rivers and streams that sustain local wildlife. The displacement of these species is likely to have ripple effects on their populations across Eastern Europe.

The Carpathian Biosphere Reserve: Located in the Carpathian Mountains, this UNESCO-protected area is home to some of Europe's largest populations of brown bears, wolves, and lynx. Military operations and deforestation caused by war have fragmented the habitats of these apex predators, reducing their hunting grounds and forcing them closer to human settlements, where they face additional risks. The reserve's ecosystems, which also support a variety of plant species unique to the region, have suffered irreversible damage.

Wetlands Along the Dnipro River: The Dnipro River and its associated wetlands are vital for aquatic species and migratory birds. Bombings and industrial spills have polluted these waters, killing fish and disrupting the food chains that sustain bird populations. The black stork, which relies on these wetlands for nesting, is now at greater risk of losing its habitat entirely. The destruction of these wetlands also reduces their capacity to filter pollutants and act as carbon sinks, adding an environmental dimension to the conflict's legacy.

Broader Implications for Ecosystems and Conservation

The loss of biodiversity during conflict is not just a local issue—it has global implications. Ecosystems are interconnected, and the decline or extinction of species in one region can disrupt food chains and ecological balances elsewhere. For instance, the destruction of Ukraine's wetlands and forests affects migratory birds that travel across continents, potentially impacting ecosystems far from the war zone.

Moreover, the degradation of protected areas undermines decades of conservation work. The resources and expertise invested in safeguarding endangered species and restoring habitats are rendered meaningless in the face of war. Rebuilding these efforts after the conflict will require significant time, funding, and international collaboration, as well as careful planning to prevent further loss of biodiversity.

The impact of war on endangered species and protected areas highlights the often-overlooked cost of conflict. The destruction of biodiversity hotspots, the threats to endangered species from habitat loss and poaching, and the specific cases from Ukraine demonstrate the urgent need for integrating conservation into conflict resolution and recovery efforts. Without immediate attention, the ecological damage caused by war could lead to irreversible losses for some of the world's most vulnerable species and ecosystems.

4.2 The Migration and Displacement of Animals Due to War Zones

War zones are not only dangerous for humans but also for the animals that inhabit these regions. Wildlife caught in conflict-affected areas faces a unique set of challenges, including habitat destruction, loud explosions, fires, and contamination of water and food sources. As a result, many animals are forced to flee their natural habitats, triggering widespread migration and displacement. These disruptions have far-reaching consequences for ecosystems, food chains, and biodiversity, with effects that last long after the conflict ends.

Forced Animal Migration: Disrupted Ecosystems and Food Chains

One of the immediate impacts of war on wildlife is the forced migration of animals. Loud explosions, fire, and the destruction of forests and wetlands compel animals to abandon their habitats in search of safety. In some cases, animals are displaced hundreds of kilometers away, moving into unfamiliar territories where they struggle to adapt to new conditions and compete with resident species for limited resources.

This forced migration disrupts ecosystems by altering the natural balance between predators, prey, and plant life. For instance, apex predators such as wolves and lynx, which require large territories for hunting, may encroach on human settlements or other ecosystems, increasing the risk of conflict with humans and further destabilizing the food chain. Prey species, such as deer or rabbits, may overpopulate certain areas when their predators are displaced, leading to overgrazing and vegetation loss.

In Ukraine, species like the Eurasian lynx and the European bison have been observed moving away from conflict zones into neighboring areas, where they face new ecological challenges. Similarly, migratory birds that rely on wetlands and forests in war-affected regions have been forced to alter their routes, potentially disrupting their breeding and feeding patterns.

Conflict Zones as Barriers to Natural Migration Patterns

War zones act as physical and psychological barriers to the natural migration of animals. Explosions, minefields, and active battlefields make it difficult for wildlife to traverse their traditional routes. Fences, trenches, and other wartime structures further fragment habitats, isolating populations and preventing genetic exchange between groups.

For migratory species, such as birds and large mammals, these barriers can have devastating consequences. Animals that are unable to complete their migrations may be cut off from vital resources, such as breeding grounds, food supplies, or water sources. In the case of birds that rely on wetlands in Ukraine for stopovers during their migratory journeys, the loss of these habitats can disrupt entire populations, affecting ecosystems far beyond the immediate war zone.

For example, the Carpathian Mountains, which serve as a critical corridor for species like bears and lynx, have been impacted by the war in Ukraine. Military activity and deforestation in these areas have created bottlenecks, preventing animals from accessing key parts of their range. This isolation increases the risk of inbreeding and weakens populations over time.

Long-Term Ecological Consequences of Displaced Wildlife Populations

The displacement of animals during conflict has long-term ecological consequences that extend beyond the immediate war zone. When species are forced to migrate into new areas, they can disrupt local ecosystems by competing with resident species for food, shelter, and space. In some cases, invasive behaviors may emerge, where displaced species become dominant in their new environments, altering the natural balance.

Additionally, the absence of species from their original habitats can lead to ecological collapse. Keystone species, such as predators or pollinators, play a crucial role in maintaining ecosystem health. Their displacement or extinction can trigger a cascade of changes that affect plant life, soil quality, and other animal populations. For example, if a predator like the wolf is removed from an ecosystem, prey species may overpopulate, leading to vegetation loss and soil degradation.

The long-term recovery of ecosystems after a conflict is also complicated by the absence of displaced wildlife. Reintroducing species to their original habitats can be challenging, particularly if those habitats have been degraded or if barriers, such as minefields, remain. Without intervention, the ecological gaps created by displaced wildlife may take decades to heal, if they recover at all.

Relevance to the Russia-Ukraine War

The ongoing war in Ukraine provides a stark example of how animal migration and displacement can affect ecosystems. The destruction of forests, wetlands, and grasslands has forced numerous species to abandon their habitats, with ripple effects throughout the region. Animals fleeing conflict zones often encounter new threats, including increased poaching, competition for resources, and the dangers of human-wildlife conflict.

For instance, migratory birds that rely on the wetlands of the Dnipro River are facing disruptions to their feeding and nesting patterns due to pollution and habitat loss. Similarly, large mammals, such as deer and bears, are being forced into areas closer to human settlements, leading to greater risks of conflict and stress for both wildlife and people.

Global Implications of War-Induced Wildlife Displacement

The displacement of wildlife during war is not just a local issue; it has global implications for biodiversity and conservation. Migratory species, which connect ecosystems across continents, are particularly vulnerable to the effects of war. Disruptions to their routes and habitats can impact ecosystems thousands of kilometers away, creating a chain reaction of environmental changes.

The lessons from Ukraine and other conflict zones highlight the need for integrating wildlife protection into wartime and post-war recovery strategies. Efforts to safeguard migration corridors, minimize habitat destruction, and monitor displaced populations are critical to mitigating the long-term ecological consequences of war. Without such measures, the impact of animal displacement will continue to undermine biodiversity and ecosystem health on a global scale.

The forced migration and displacement of animals during war illustrate the hidden cost of conflict on wildlife and ecosystems. The disruptions to food chains, migration patterns, and ecological balances create challenges that extend far beyond the immediate war zone. Addressing these issues requires a comprehensive approach that prioritizes habitat restoration, wildlife protection, and sustainable recovery to ensure that both humans and animals can coexist in a healthier, more resilient environment.

4.3 Collateral Damage to Ecosystems and Food Webs

War exerts a devastating toll on ecosystems and the intricate food webs that sustain them. The environmental destruction caused by conflict disrupts water sources, fragments habitats, and destabilizes entire food chains, creating a cascade of effects that ripple through ecosystems for years, if not decades. The Russia-Ukraine war has exemplified how military activity can undermine ecological stability, with consequences for biodiversity, natural resource sustainability, and regional environmental health.

Damage to Water Sources and Their Impact on Aquatic Species

Water systems are among the most vulnerable components of ecosystems during conflict. Bombings, chemical spills, and infrastructure destruction often contaminate rivers, lakes, and wetlands, affecting aquatic species and the ecosystems they support. When pollutants such as oil, heavy metals, or chemical residues enter water sources, they degrade water quality, disrupt food chains, and create conditions that are inhospitable for fish, amphibians, and aquatic plants.

In Ukraine, the Dnipro River, one of Europe's major waterways, has been severely affected by the war. Bombing near industrial facilities and fuel depots has led to toxic runoff entering the river, killing fish and other aquatic organisms. Polluted water diminishes oxygen levels, causing fish populations to suffocate and disrupting the delicate balance of aquatic ecosystems. Migratory species that depend on these waters as feeding or breeding grounds face challenges that extend far beyond the immediate region.

The contamination of water sources also impacts terrestrial animals and humans that rely on these waters for drinking and sustenance. As aquatic species decline, so do the predators and scavengers that depend on them, creating a ripple effect that destabilizes the broader ecosystem. Moreover, the loss of aquatic biodiversity reduces the ability of water systems to purify themselves, leading to long-term environmental degradation.

Fragmentation of Habitats Caused by Military Activity

Military operations cause significant habitat fragmentation, breaking once-contiguous ecosystems into isolated patches. This fragmentation occurs through the construction of trenches, barriers, and roads, as well as through the destruction of forests, wetlands, and grasslands. Fragmented habitats disrupt the movement of species, limiting their access to food, mates, and breeding grounds.

In Ukraine, forests and grasslands have been fragmented by bombing campaigns, troop movements, and the establishment of defensive positions. Animals that rely on large, undisturbed territories, such as deer, wolves, and lynx, are particularly affected. Fragmentation forces these species to travel greater distances to find resources, increasing their risk of encountering human settlements or becoming prey to other predators.

Habitat fragmentation also isolates populations, reducing genetic diversity and making species more vulnerable to disease and environmental changes. For example, amphibians in wetlands impacted by the war may face challenges finding suitable breeding areas, leading to population declines. Similarly, plant species that rely on pollinators may suffer as fragmented landscapes make it harder for insects to traverse large areas.

The effects of habitat fragmentation extend to the broader ecosystem, disrupting natural processes such as pollination, seed dispersal, and predator-prey interactions. Over time, these disruptions can lead to the collapse of local ecosystems, with cascading impacts on biodiversity and ecological health.

The Cascading Effects of War on Food Chains and Ecosystem Stability

War-induced damage to ecosystems has far-reaching consequences for food chains and overall ecosystem stability. The destruction of keystone species—those that play a critical role in maintaining ecological balance—can trigger a chain reaction that affects multiple levels of the food web. For instance, the loss of predators such as wolves or raptors allows prey populations to overpopulate, leading to overgrazing or depletion of vegetation.

Similarly, the decline of primary producers, such as aquatic plants or forest vegetation, disrupts the base of the food chain, affecting herbivores and the predators that depend on them. In Ukraine, wetlands and forests that have been destroyed by military activity no longer support the same levels of biodiversity, reducing the availability of food and shelter for species at all levels of the food chain.

Aquatic food chains are particularly vulnerable to the cascading effects of war. The contamination of rivers and lakes impacts phytoplankton and algae, the primary producers in aquatic ecosystems. As these organisms decline, so do the fish and invertebrates that feed on them, ultimately affecting birds and mammals that rely on aquatic species as part of their diet.

Ecosystem stability is further undermined by the introduction of invasive species that take advantage of the disrupted environment. These species often outcompete native ones, altering the structure of the food web and reducing biodiversity. For example, areas in Ukraine where native vegetation has been destroyed may become susceptible to invasive plants that thrive in disturbed soil, crowding out local flora and disrupting herbivore populations.

Broader Implications of Ecosystem Disruption

The damage to water sources, habitat fragmentation, and cascading effects on food chains have profound implications for the long-term health of ecosystems. Once an ecosystem is destabilized, it can take years or even decades to recover fully, and in some cases, the damage may be irreversible. The loss of biodiversity, reduced ecosystem services, and increased vulnerability to climate change further compound the challenges of post-war recovery.

In the context of the Russia-Ukraine war, these ecological impacts highlight the need for comprehensive environmental restoration efforts that address the full scope of ecosystem damage. Rehabilitating water sources, reconnecting fragmented habitats, and reintroducing displaced species are critical steps toward rebuilding ecological stability. Without these measures, the cascading effects of war on ecosystems will continue to undermine biodiversity and natural resource sustainability.

War's collateral damage to ecosystems and food webs is a stark reminder of how deeply interconnected human conflicts are with the natural world. By disrupting water sources, fragmenting habitats, and destabilizing food chains, wars create environmental crises that persist long after the fighting ends. Addressing these impacts requires not only immediate action but also long-term commitments to conservation and sustainable recovery, ensuring that ecosystems can heal and thrive once more.

4.4 Efforts by Conservation Groups Amidst Conflict

While war wreaks havoc on wildlife and ecosystems, conservation groups play a crucial role in mitigating the damage, often working under extraordinary circumstances. From rescuing animals trapped in conflict zones to safeguarding biodiversity and restoring habitats, these organizations strive to minimize the long-term environmental toll of war. The Russia-Ukraine war has prompted both local and international conservation groups to step up their efforts, often collaborating to address immediate and long-term challenges.

Emergency Rescue Operations for Wildlife in War Zones

One of the most immediate priorities for conservation groups during conflicts is rescuing animals trapped in war zones. The chaos of war leaves countless animals injured, displaced, or stranded, often with no access to food or safe habitats. Conservation organizations mobilize quickly to evacuate wildlife from high-risk areas, provide medical care, and relocate them to safer environments.

In Ukraine, several animal rescue operations have been launched in regions heavily affected by the war. Zoos and wildlife parks, often located near urban centers, have been particularly vulnerable to bombings and shelling. Organizations such as the Ukrainian Nature Conservation Group (UNCG) and international partners like the World Wildlife Fund (WWF) have worked tirelessly to transport animals to safer locations, including neighboring countries like Poland and Romania.

For instance, when the Mykolaiv Zoo came under threat, emergency teams coordinated the evacuation of rare and endangered species, including large mammals and birds. These operations require significant logistical planning, as moving stressed or injured animals through active war zones poses unique challenges. Despite these obstacles, conservationists have successfully rescued numerous animals, ensuring their survival amid the chaos.

Protecting Biodiversity Under the Threat of Conflict

Conservation efforts during war go beyond immediate rescue operations, focusing on protecting biodiversity and ecosystems from further damage. This involves monitoring endangered species, safeguarding critical habitats, and preventing illegal activities such as poaching and deforestation, which often escalate during conflicts.

In Ukraine, conservationists have established monitoring programs to track the movement of endangered species like the European bison, lynx, and black stork. Using GPS tracking and remote sensing technologies, they collect data on animal populations and habitat conditions, even in areas that are difficult or unsafe to access. This information is vital for assessing the impact of the conflict on biodiversity and planning future restoration efforts.

Additionally, conservation groups work to combat the rise in poaching that often accompanies war. The absence of law enforcement and the desperation of local communities can lead to increased hunting of wildlife for food or profit. Organizations like the Ukrainian Society for the Protection of Birds have been raising awareness about the importance of preserving biodiversity, even during difficult times, and are working with local authorities to prevent illegal activities in protected areas.

Collaboration Between Local and Global Organizations for Conservation

Collaboration between local and international organizations is essential for effective conservation efforts in conflict zones. Local groups bring invaluable knowledge of the region's ecosystems and wildlife, while global organizations provide resources, expertise, and international advocacy to support these efforts.

The Russia-Ukraine war has seen a surge in international partnerships aimed at addressing the environmental consequences of the conflict. The International Union for Conservation of Nature (IUCN), for example, has partnered with Ukrainian conservation groups to fund emergency projects and develop strategies for post-war ecological recovery. These collaborations often focus on sharing resources, such as satellite imagery for habitat monitoring or funding for wildlife rescue operations.

Global organizations also play a key role in bringing attention to the environmental impacts of war on the international stage. By documenting the ecological damage caused by the conflict, they push for the inclusion of environmental considerations in peace negotiations and recovery plans. Initiatives like the Environmental Peacebuilding Association advocate for the integration of conservation into conflict resolution, emphasizing the importance of protecting ecosystems as part of rebuilding efforts.

In addition to global partnerships, regional networks have proven instrumental in supporting conservation during the war. Neighboring countries have provided sanctuaries for displaced animals and facilitated cross-border collaboration on conservation projects. For example, Poland and Romania have assisted in hosting wildlife evacuated from Ukraine, ensuring that species threatened by the conflict can survive and eventually be reintroduced to their natural habitats.

Overcoming Challenges in Conservation During War

Conservation groups face immense challenges in their efforts to protect wildlife and ecosystems during conflicts. Limited access to war zones, the risk of personal harm, and the destruction of infrastructure complicate their work. Additionally, securing funding for conservation projects can be difficult, as international attention often prioritizes humanitarian aid over environmental concerns.

Despite these obstacles, conservationists have demonstrated remarkable resilience and innovation. They have used technology to monitor ecosystems remotely, engaged local communities in conservation efforts, and adapted their strategies to respond to the dynamic challenges of war. These efforts highlight the critical role of conservation in preserving biodiversity and mitigating the long-term environmental impacts of conflict.

The work of conservation groups amidst the Russia-Ukraine war underscores the importance of protecting wildlife and ecosystems even in the face of human tragedy. By conducting emergency rescues, safeguarding biodiversity, and fostering international collaboration, these organizations are laying the groundwork for ecological recovery and sustainability. Their efforts not only ensure the survival of individual species but also contribute to the broader goal of preserving the natural world for future generations.

4.5 Post-War Recovery of Habitats and Wildlife

The environmental destruction caused by war leaves ecosystems and wildlife in a fragile state, requiring significant efforts for recovery and restoration. Post-war recovery of habitats and wildlife is not just about repairing visible damage; it also involves rebuilding the intricate ecological balances that sustain biodiversity and natural resources. With strategic planning, international support, and sustainable initiatives, war-torn regions can recover their ecological vitality. This section explores the key strategies for habitat restoration and species recovery, the role of international funding, and the integration of biodiversity rebuilding into post-conflict environmental initiatives.

Strategies for Habitat Restoration and Species Recovery

Restoring habitats and recovering wildlife populations in post-war regions is a complex process that demands both immediate and long-term actions. Key strategies include:

Rehabilitating Damaged Ecosystems

The first step in habitat restoration involves clearing debris, contaminants, and unexploded ordnance (UXO) from affected areas. For forests, reforestation efforts using native tree species can help restore lost vegetation, while wetlands and riverbanks require measures to stabilize soil and prevent further erosion. Specialized ecological assessments are crucial for identifying areas in urgent need of intervention and planning tailored restoration projects.

For example, in Ukraine, reforestation programs could target the Carpathian Biosphere Reserve and Polissya Biosphere Reserve, which have been heavily impacted by military activity. Wetland restoration along the

Dnipro River would also help rebuild critical aquatic ecosystems that support migratory birds and aquatic life.

Protecting and Reintroducing Wildlife

Species recovery programs are essential to repopulate areas that have lost key wildlife populations. This may involve reintroducing species that were displaced or whose populations have dwindled due to habitat destruction. Careful planning ensures that reintroduction efforts consider genetic diversity, prey availability, and habitat suitability.

In Ukraine, species like the Eurasian lynx, European bison, and black stork may require reintroduction into restored habitats. Temporary wildlife sanctuaries, established during the war, can serve as a foundation for these programs by providing a controlled environment for breeding and rehabilitation before reintroduction.

Combating Invasive Species

The disruption caused by war often allows invasive species to take root, outcompeting native flora and fauna. Post-war recovery efforts must include measures to identify and remove invasive species to restore the natural balance of ecosystems.

Monitoring and Research

Continuous monitoring of restored habitats and wildlife populations is critical to ensure the success of recovery efforts. Research into the long-term impacts of war on ecosystems can inform adaptive management strategies, helping to address emerging challenges such as climate change and ongoing environmental degradation.

International Funding and Support for Ecological Recovery

Post-war ecological recovery often requires financial and technical support beyond the capacity of the affected nation. International funding and collaboration play a pivotal role in addressing the extensive damage caused by war. Key elements of international support include:

Funding for Restoration Projects

Global organizations like the United Nations Development Programme (UNDP), World Wildlife Fund (WWF), and Green Climate Fund can provide the financial resources needed to launch large-scale restoration projects. These funds can be directed toward reforestation, wetland recovery, and wildlife protection initiatives.

In Ukraine, the European Union and international donors have already expressed interest in supporting environmental recovery efforts. Allocating part of reconstruction aid to ecological projects can ensure that nature is prioritized alongside infrastructure rebuilding.

Capacity Building and Expertise

International conservation organizations can offer technical expertise in areas such as habitat restoration, species recovery, and environmental monitoring. Training local conservationists and communities ensures that recovery efforts are sustainable and adapted to local conditions.

Cross-Border Collaboration

Ecosystems often span national boundaries, making cross-border collaboration essential for effective recovery. Neighboring countries can assist by hosting displaced wildlife, sharing resources, and working jointly on transboundary conservation projects. For instance, Ukraine's Carpathian Mountain ecosystem, which extends into Romania, Slovakia, and Poland, could benefit from a regional recovery strategy.

Environmental Advocacy and Diplomacy

International advocacy can help secure long-term commitments to ecological recovery. By highlighting the global significance of Ukraine's biodiversity and ecosystems, conservation groups can mobilize resources and foster international solidarity for environmental initiatives.

Rebuilding Biodiversity as Part of Post-Conflict Environmental Initiatives

Integrating biodiversity recovery into broader post-conflict environmental initiatives ensures that ecosystems are restored alongside human infrastructure and societal stability. Key components include:

Aligning Recovery with Sustainability Goals: Rebuilding efforts should align with global sustainability goals, such as the United Nations' Sustainable Development Goals (SDGs). Incorporating renewable energy, sustainable agriculture, and green infrastructure into recovery plans can reduce the environmental footprint of reconstruction while fostering long-term ecological health.

Creating Protected Areas and Wildlife Corridors: Expanding protected areas and establishing wildlife corridors can help connect fragmented habitats, allowing species to move freely and rebuild populations. These corridors are particularly important for migratory species and apex predators that require large territories.

For Ukraine, enhancing protected areas in the Carpathians and reestablishing wetlands along the Dnipro River could serve as cornerstone projects for biodiversity recovery. Collaborative frameworks with neighboring countries could also help protect transboundary ecosystems.

Engaging Local Communities: Local communities play a critical role in the success of biodiversity recovery. Involving residents in reforestation, wildlife monitoring, and sustainable land management creates a sense of ownership and ensures long-term stewardship of restored habitats. Providing incentives, such as eco-tourism opportunities and sustainable livelihoods, can further strengthen community support.

Leveraging Technology and Innovation: Advanced technologies, such as satellite imagery, GIS mapping, and drones, can aid in monitoring ecosystems and tracking species recovery. These tools allow conservationists to assess progress, identify emerging threats, and adjust strategies as needed.

Institutionalizing Environmental Recovery: Governments must prioritize environmental recovery in post-conflict policies and budgets. Establishing dedicated agencies or task forces to oversee ecological restoration ensures that biodiversity recovery remains a central focus of reconstruction efforts.

Relevance to the Russia-Ukraine War

For Ukraine, post-war recovery presents a unique opportunity to rebuild its ecosystems and wildlife with sustainability in mind. The nation's rich biodiversity and natural resources make ecological recovery not only a local necessity but also a global priority. By integrating habitat restoration and wildlife recovery into its broader reconstruction efforts, Ukraine can emerge as a leader in post-conflict environmental recovery.

International collaboration and funding will be critical in achieving these goals. With the right support, Ukraine can restore its natural landscapes, safeguard its wildlife, and create a resilient ecological foundation for future generations. These efforts will not only benefit Ukraine but also contribute to global biodiversity and climate stability.

The post-war recovery of habitats and wildlife is a long and challenging process, but it is essential for restoring the ecological balance disrupted by conflict. By employing strategic restoration initiatives, leveraging international support, and prioritizing biodiversity in reconstruction efforts, war-torn regions can transform environmental devastation into an opportunity for renewal and sustainability. For Ukraine and other nations recovering from conflict, these efforts hold the promise of a brighter and more resilient ecological future.

Conclusion: The Forgotten Victims: Wildlife and Habitats

The impact of war on wildlife and habitats is a stark reminder of how deeply human conflicts affect the natural world. From the destruction of biodiversity hotspots to the displacement of animals and the collapse of food webs, war leaves a profound and lasting imprint on ecosystems. As explored in this chapter, the Russia-Ukraine conflict has turned forests, wetlands, and protected areas into battlegrounds, disrupting the delicate balance of nature and threatening the survival of countless species.

Conservation groups have stepped up amidst the chaos, launching rescue operations, protecting biodiversity, and fostering international collaborations to address immediate and long-term challenges. However, the road to recovery for habitats and wildlife in war-torn regions is a complex and enduring process. Strategies for habitat restoration, species reintroduction, and post-conflict ecological initiatives must prioritize sustainability, align with global climate goals, and engage local communities.

The plight of Ukraine's wildlife is a microcosm of the global consequences of war on nature. It highlights the urgent need to integrate environmental considerations into both conflict resolution and post-war rebuilding efforts. By recognizing and addressing the ecological toll of conflict, humanity can take a critical step toward preserving the interconnected systems that sustain life on Earth.

As Ukraine rebuilds, the opportunity to restore its ecosystems and protect its biodiversity offers hope for a more resilient and harmonious coexistence with nature. This chapter calls for a commitment to ensuring that the forgotten victims of war—wildlife and habitats—are given the attention and care they need to recover and thrive once more.

Chapter 5: Rebuilding or Restoring?

The end of war marks the beginning of another battle—the fight to rebuild what has been lost. For Ukraine, this means not only reconstructing cities, infrastructure, and industries but also restoring its natural environment, which has suffered extensive damage. The question, however, is not simply about rebuilding but whether this recovery will prioritize sustainability and ecological health or focus solely on economic growth at the expense of the environment.

Environmental recovery is a complex task that requires balancing immediate needs with long-term goals. From clearing landmines and toxic remnants of war to restoring ecosystems and biodiversity, the process must address the multifaceted impacts of conflict on nature. This chapter explores the critical challenges Ukraine faces in rebuilding its environmental and ecological systems, the role of international organizations and treaties, and the strategies needed to ensure that restoration, not just reconstruction, guides the country's path forward.

5.1 Environmental Recovery in Post-War Ukraine

The environmental recovery of Ukraine after the war presents a unique opportunity to rebuild the nation with sustainability and ecological health in mind. However, the path forward is fraught with challenges, as the ecological aftermath of war has left vast stretches of land and water systems damaged. The immediate need for economic recovery often clashes with the long-term goal of restoring ecosystems, posing a difficult question: how can Ukraine balance these priorities while avoiding further harm to its environment? This section delves into the key challenges of environmental recovery, the importance of prioritizing ecosystem restoration, and strategies to align economic growth with sustainability.

The Ecological Aftermath of War: A Summary of Challenges

The environmental toll of the Russia-Ukraine war is profound and multifaceted, impacting air, soil, water, and biodiversity across the country. Key challenges include:

Contaminated Land and Water: The widespread use of explosives, bombings, and chemical spills has left soil and water sources heavily polluted with toxic substances, including heavy metals, hydrocarbons, and other hazardous materials. Agricultural lands, vital to Ukraine's role as a global food supplier, are now unusable in many areas due to contamination or the presence of unexploded ordnance (UXO).

Destroyed Ecosystems: Forests, wetlands, and biodiversity hotspots have been reduced to battlefields, leading to habitat destruction, displacement of wildlife, and disruption of ecological processes. These areas, which once served as carbon sinks and supported local communities, now face long-term degradation.

Loss of Biodiversity: Endangered species such as the Eurasian lynx, European bison, and black stork face heightened threats as their habitats are destroyed or fragmented. The loss of keystone species could destabilize ecosystems further, leading to cascading ecological consequences.

Scarcity of Resources for Restoration: With much of the country's financial resources diverted to rebuilding critical infrastructure, the allocation of funds for environmental restoration may become a secondary concern. This could delay recovery and exacerbate long-term environmental damage.

Prioritizing Ecosystem Restoration Over Rapid Industrial Rebuilding

The urgency to rebuild Ukraine's economy and infrastructure after the war is understandable, but prioritizing rapid industrialization without considering ecological consequences could lead to irreversible environmental harm. Instead, Ukraine has an opportunity to adopt a restoration-first approach that integrates sustainability into its recovery efforts. Key principles of this approach include:

Restoring Natural Ecosystems: Ecosystem restoration must take precedence in areas where environmental damage threatens long-term ecological stability. Reforestation programs, wetland rehabilitation, and soil decontamination are vital steps to restoring the natural balance of affected regions. These efforts will not only support biodiversity recovery but also enhance Ukraine's resilience to climate change by restoring carbon sinks and natural water filtration systems.

Using Nature-Based Solutions: Nature-based solutions, such as restoring floodplains, planting trees, and protecting existing ecosystems, can provide cost-effective alternatives to traditional infrastructure rebuilding. For example, restoring wetlands along the Dnipro River can mitigate flood risks while supporting biodiversity and water quality.

Slowing Industrial Expansion in Vulnerable Areas: To avoid further damage, Ukraine must regulate industrial activity in ecologically sensitive areas, such as forests and wetlands. This includes implementing stricter environmental impact assessments for reconstruction projects and encouraging the development of green industries that minimize their ecological footprint.

Leveraging Renewable Energy in Reconstruction: Rebuilding Ukraine's energy systems presents an opportunity to shift from fossil fuels to renewable energy sources, such as wind, solar, and bioenergy. Renewable energy infrastructure can reduce carbon emissions while providing a sustainable foundation for economic growth.

Balancing Economic Recovery with Environmental Sustainability

Achieving a balance between economic recovery and environmental sustainability requires careful planning, investment, and collaboration. Key strategies include:

Integrating Environmental Goals into National Recovery Plans: Ukraine's recovery strategy must explicitly include environmental restoration as a core component, ensuring that ecological health is prioritized alongside economic growth. This involves setting clear targets for reforestation, biodiversity recovery, and pollution cleanup within the broader reconstruction agenda.

Promoting Sustainable Agriculture: Agriculture, a cornerstone of Ukraine's economy, must transition to sustainable practices to address soil degradation and contamination. Regenerative agriculture techniques, such as crop rotation, reduced tillage, and organic farming, can restore soil health and increase productivity without relying on harmful chemicals.

Engaging Local Communities in Restoration Efforts: Local communities have a critical role to play in environmental recovery. Providing incentives for farmers and landowners to participate in reforestation and conservation projects can accelerate progress while creating jobs and fostering a sense of stewardship for natural resources.

Collaborating with International Partners: Ukraine's environmental recovery will require significant international support in the form of funding, expertise, and technology. Partnerships with global conservation organizations, development agencies, and neighboring countries can help Ukraine achieve its sustainability goals while rebuilding its economy.

Building Green Infrastructure: Infrastructure rebuilding offers a chance to integrate sustainable practices, such as energy-efficient buildings, green urban spaces, and public transportation systems. These measures can reduce environmental impacts while improving the quality of life for Ukrainian citizens.

Relevance to Ukraine's Future

By prioritizing ecosystem restoration and sustainable practices, Ukraine can turn the challenge of environmental recovery into an opportunity for long-term growth and resilience. Aligning its recovery efforts with global environmental goals, such as the United Nations' Sustainable Development Goals (SDGs) and the Paris Agreement, will position Ukraine as a leader in post-conflict sustainability.

This approach will also strengthen Ukraine's natural resource base, ensuring that its forests, wetlands, and agricultural lands can continue to support both local communities and global markets. While the road to recovery will be long and complex, placing the environment at the center of Ukraine's rebuilding strategy is essential for securing a healthier, more sustainable future.

5.2 Challenges of Clearing Landmines and Unexploded Ordnance (UXO)

Landmines and unexploded ordnance (UXO) are among the most devastating legacies of war, posing long-term risks to both human and environmental safety. Ukraine, with vast stretches of land contaminated by UXO due to the ongoing conflict, faces a monumental challenge in safely clearing these remnants of war. The presence of landmines and UXO not only threatens lives but also hinders efforts to restore ecosystems, reclaim arable land, and rebuild the country's economy. This section explores the environmental and ecological risks of landmines, methods for safe demining, and strategies for restoring land and ecosystems after clearance.

The Environmental and Ecological Risks of Landmines

Soil Degradation and Contamination: Landmines and UXO release toxic chemicals into the soil as they degrade over time. Explosives such as TNT and RDX, along with heavy metals from mine casings, leach into the ground, contaminating the soil and making it unsuitable for agriculture. These pollutants can persist for decades, posing a long-term threat to soil fertility and the safety of crops grown in affected areas.

Hindrance to Wildlife and Habitat Recovery: Landmines disrupt wildlife by creating barriers to movement and making habitats unsafe. Animals, unaware of the danger, can trigger landmines, leading to injuries or death. This not only impacts individual species but also destabilizes ecosystems by removing key players in the food chain. For example, predators such as wolves or lynx may lose access to hunting grounds, while herbivores like deer may face limited grazing areas.

Damage to Water Systems: Explosions from UXO can contaminate nearby water sources with soil and chemical debris, affecting aquatic ecosystems and water quality. The presence of UXO in rivers and wetlands further complicates clearance efforts, leaving these critical habitats in a state of prolonged degradation.

Obstacle to Land Reclamation: Contaminated land cannot be safely used for agriculture, conservation, or infrastructure development until it is cleared of landmines and UXO. This delays efforts to restore arable land and rebuild ecosystems, leaving large swathes of Ukraine's natural and agricultural landscapes inaccessible.

Methods and Technologies for Safe Demining in Ukraine

Demining is a complex, time-consuming, and resource-intensive process that requires a combination of traditional methods and modern technologies. In Ukraine, the scale of the problem demands innovative solutions and international collaboration to ensure safety and efficiency.

Manual Demining: Traditional manual demining remains the most accurate method for identifying and safely removing landmines and UXO. Trained personnel use metal detectors, prodding tools, and protective equipment to locate and neutralize explosives. However, this method is labor-intensive and slow, particularly in regions with high mine density.

Mechanical Clearance: Mechanical demining uses specialized vehicles equipped with flails, rollers, or plows to clear mines from the surface of the land. These machines are effective in clearing large areas quickly but can be limited by terrain, such as forests, wetlands, or rocky regions, where maneuverability is restricted.

Explosive Detection Dogs (EDDs): Dogs trained to detect the scent of explosives play a critical role in demining operations. Their ability to cover large areas quickly and with high accuracy makes them invaluable in identifying minefields. In Ukraine, organizations such as the Halo Trust have been deploying EDDs to assist with demining efforts.

Drone and Robotics Technology: Modern advancements in robotics and drone technology are revolutionizing demining. Drones equipped with ground-penetrating radar and thermal imaging can survey large areas, identifying potential minefields without putting human lives at risk. Similarly, robots with precision tools can neutralize UXO in inaccessible or high-risk areas, such as wetlands or forests.

Community Engagement and Reporting: Local communities play a key role in identifying areas contaminated with landmines and UXO.

Education and awareness campaigns encourage residents to report suspicious objects and avoid unsafe areas, reducing the risk of accidents and aiding in mapping contaminated regions.

Restoring Arable Land and Ecosystems After Landmine Clearance

Once landmines and UXO are removed, the process of restoring arable land and ecosystems begins. This requires a combination of soil rehabilitation, ecosystem restoration, and sustainable land management practices.

Soil Rehabilitation: Land that has been contaminated by explosives often requires soil testing and remediation to restore its fertility. Techniques such as bioremediation, which uses plants or microorganisms to break down pollutants, can help detoxify the soil. Adding organic matter and nutrients can further enhance soil health, making it suitable for agriculture once again.

Reforestation and Habitat Restoration: Clearing landmines paves the way for reforestation and habitat recovery. Native tree and plant species can be reintroduced to stabilize the soil, create shelter for wildlife, and rebuild ecosystems. Wetlands and riverbanks, often heavily impacted by UXO, may require additional interventions, such as water filtration systems or sediment removal, to restore their ecological functions.

Reclaiming Agricultural Land: Restoring arable land involves working closely with local farmers to reintroduce sustainable agricultural practices. Crop rotation, reduced tillage, and organic farming methods can help regenerate soil health while minimizing environmental impact. Training programs and financial support for farmers can accelerate the recovery process and reintegrate cleared land into the national economy.

Long-Term Monitoring and Maintenance: Even after clearance, affected areas require ongoing monitoring to ensure safety and ecological health. Regular inspections, community reporting systems, and technological advancements in land surveillance can help maintain cleared areas and prevent the reintroduction of hazardous materials.

Integrating Conservation Goals: Post-clearance efforts should align with broader conservation goals to protect biodiversity and restore ecosystems. For instance, rewilding initiatives in cleared areas can help reintroduce displaced wildlife and rebuild ecological networks, ensuring long-term stability and resilience.

Relevance to Ukraine's Environmental Recovery

The challenge of clearing landmines and UXO is one of the most significant barriers to Ukraine's environmental and economic recovery. Addressing this issue requires a coordinated effort involving the Ukrainian government, international organizations, and local communities. By leveraging modern technologies, fostering collaboration, and integrating environmental restoration into demining operations, Ukraine can turn cleared land into a foundation for sustainable development.

Landmine clearance is not just a safety imperative—it is a crucial step toward reclaiming Ukraine's natural heritage and rebuilding its role as an agricultural and ecological powerhouse. Success in this endeavor will demonstrate the resilience of both the nation and its environment in the face of war's devastating legacy.

5.3 Sustainable Reconstruction: Avoiding Further Environmental Harm

Post-war reconstruction is an opportunity to rebuild not just infrastructure but also a vision for a sustainable future. However, prioritizing rapid industrialization and economic recovery without considering ecological health can lead to irreversible environmental damage. For Ukraine, balancing the urgency of rebuilding with the need for sustainability is critical. By incorporating green technologies and learning from past conflicts, Ukraine can avoid repeating mistakes and set an example of how to align development with environmental preservation.

Risks of Prioritizing Rapid Industrialization Over Ecological Health

Rebuilding a nation after war often emphasizes economic recovery, infrastructure repair, and industrial development. While these are essential for restoring normalcy, prioritizing them at the expense of ecological health poses significant risks:

Environmental Degradation: Rapid industrialization often involves extensive land use, resource extraction, and pollution. Without proper planning, these activities can further degrade war-damaged ecosystems, exacerbating soil erosion, deforestation, and water contamination.

Loss of Biodiversity: Accelerated construction projects can encroach on natural habitats, leading to further fragmentation and the loss of biodiversity. Wildlife that survived the war may face new threats from habitat destruction caused by infrastructure development.

Increased Carbon Emissions: Traditional construction methods and energy-intensive industries rely heavily on fossil fuels, leading to a spike in greenhouse gas emissions. In a country like Ukraine, which already faces environmental challenges from the conflict, unchecked industrialization could hinder its progress toward global climate goals.

Missed Opportunities for Sustainable Growth: Focusing solely on short-term economic recovery can lock a nation into outdated, carbon-intensive systems. This limits the potential for integrating renewable energy, green technologies, and sustainable urban planning, which could create long-term economic and environmental benefits.

Incorporating Green Technologies into Infrastructure Rebuilding

To avoid further environmental harm, Ukraine has a unique opportunity to incorporate green technologies and sustainable practices into its reconstruction efforts. Key areas where this can be achieved include:

Renewable Energy Integration: Rebuilding Ukraine's energy infrastructure provides an opportunity to transition from fossil fuels to renewable energy sources such as solar, wind, and bioenergy. For instance, solar farms can be established in areas unsuitable for agriculture due to landmine contamination, turning otherwise unusable land into a productive energy resource.

Energy-Efficient Buildings: Incorporating green building technologies, such as energy-efficient designs, insulation, and smart grids, can reduce energy consumption in residential, commercial, and public buildings. Using materials with lower environmental footprints, like recycled steel and sustainably sourced timber, further minimizes the ecological impact of construction.

Sustainable Transportation Systems: Rebuilding Ukraine's transportation infrastructure offers a chance to promote sustainable mobility solutions, such as electric public transport, cycling lanes, and energy-efficient rail systems. Modernizing transportation networks with eco-friendly technologies can reduce emissions and improve urban living standards.

Reforestation and Urban Green Spaces: Urban reconstruction plans should include reforestation projects and the development of green spaces within cities. These areas not only improve air quality and provide climate resilience but also enhance the quality of life for residents, fostering social cohesion in post-war communities.

Smart Water Management: Restoring water systems damaged by the war requires sustainable approaches, such as rainwater harvesting, wastewater recycling, and the use of constructed wetlands for water purification. These solutions support both human needs and ecosystem health, ensuring a balanced approach to resource use.

Case Studies of Sustainable Post-War Reconstruction from Other Conflicts

Learning from past conflicts can provide valuable insights into how sustainable practices can be integrated into post-war rebuilding. Several examples highlight the importance of balancing recovery with ecological preservation:

Rwanda: Restoring Ecosystems Post-Genocide: Following the 1994 genocide, Rwanda implemented extensive reforestation and conservation programs as part of its national recovery strategy. The government established protected areas and reintroduced native tree species, resulting in significant improvements in biodiversity and climate resilience. Today, Rwanda is a leader in sustainable development, with a focus on renewable energy and eco-tourism.

Bosnia and Herzegovina: Sustainable Energy Development: After the Balkan wars of the 1990s, Bosnia and Herzegovina faced the challenge of rebuilding its energy sector. By investing in hydropower and renewable energy projects, the country reduced its reliance on coal while creating jobs in the green energy sector. This approach demonstrated how integrating sustainability into recovery can drive economic growth while reducing environmental harm.

Germany Post-World War II: Green Urban Planning: The post-World War II rebuilding of German cities like Frankfurt and Berlin emphasized green urban planning and the development of efficient public transportation systems. Today, these cities serve as examples of how sustainable urban development can create livable, resilient, and environmentally friendly urban spaces.

Vietnam: Mangrove Restoration Post-War: The environmental damage caused by chemical defoliants during the Vietnam War devastated mangrove ecosystems. In the years following the conflict, Vietnam implemented large-scale mangrove reforestation programs, restoring critical habitats and enhancing coastal resilience against climate change. These efforts have benefited both biodiversity and local communities that rely on mangroves for livelihoods.

Relevance to Ukraine's Reconstruction Efforts

Ukraine can draw on these examples to develop its own sustainable reconstruction strategy. By prioritizing green technologies and aligning rebuilding efforts with global sustainability goals, Ukraine has the opportunity to position itself as a leader in eco-friendly development. This approach not only mitigates the long-term environmental impact of the war but also supports economic resilience by creating jobs in green industries and attracting international funding.

Sustainable reconstruction must be seen not as a secondary concern but as a cornerstone of Ukraine's recovery. By investing in renewable energy, modern infrastructure, and ecological restoration, Ukraine can rebuild stronger, greener, and more resilient for future generations.

Interview with Dr. Maria Novak, Urban Planner Specializing in Sustainable Reconstruction

Dr. Maria Novak is an urban planner with extensive experience in designing sustainable and resilient cities in post-conflict regions. She has worked on several international recovery projects, focusing on integrating eco-friendly practices into rebuilding efforts to ensure long-term environmental and societal stability.

Q: Dr. Novak, thank you for joining us. Can you start by explaining why sustainable reconstruction is so critical in post-war recovery efforts?

A: Thank you for having me. Sustainable reconstruction is vital because post-war recovery isn't just about rebuilding what was lost—it's about building better, stronger, and more resilient systems that can withstand future challenges. War leaves behind not only physical destruction but also significant environmental damage. If reconstruction efforts don't consider sustainability, we risk exacerbating these issues, creating vulnerabilities for future generations.

For example, rebuilding energy infrastructure with fossil fuels may restore power quickly, but it locks communities into unsustainable energy dependencies. Similarly, rebuilding housing without considering energy efficiency can lead to long-term resource wastage and increased carbon footprints. Sustainable reconstruction is an opportunity to address these challenges while promoting economic growth and environmental resilience.

Q: What are some key principles of sustainable reconstruction, and how can they be applied in conflict zones like Ukraine?

A: The key principles of sustainable reconstruction include resilience, resource efficiency, and inclusivity. Let me break these down:

1. **Resilience**: Infrastructure should be designed to withstand both physical and cyber threats. For example:

✓ Energy grids can incorporate renewable sources like solar and wind to reduce dependence on centralized, vulnerable systems.

✓ Buildings can be designed with disaster-resilient materials and structures that can endure natural and human-made threats.

1. **Resource Efficiency**: Recovery efforts should prioritize minimizing resource consumption and waste. This includes:

✓ Using recycled or locally sourced materials to reduce the environmental footprint.

✓ Incorporating water-saving technologies like smart irrigation systems and graywater recycling in urban designs.

1. **Inclusivity**: Reconstruction should involve local communities to ensure that their needs and priorities are met. This means:

✓ Designing spaces that are accessible and promote social cohesion.

✓ Training local workers to implement sustainable practices, creating jobs and fostering ownership of the rebuilding process.

In Ukraine, these principles could translate into projects like rebuilding cities with green roofs to manage stormwater, integrating renewable energy into the grid, and designing public spaces that double as natural flood barriers.

Q: What are the biggest challenges in implementing sustainable reconstruction in post-war environments?

A: The challenges are significant, but they're not insurmountable. One of the biggest hurdles is **resource availability**. Conflict zones often face shortages of funding, materials, and skilled labor, making it difficult to prioritize sustainable practices over immediate needs.

Another challenge is **coordination among stakeholders**. Post-war recovery typically involves multiple actors—governments, NGOs, private companies, and local communities—each with their own priorities and timelines. Aligning these interests around sustainability requires strong leadership and clear frameworks.

Finally, there's the **issue of urgency**. After a conflict, the focus is often on rapid reconstruction to restore basic services and housing. While this urgency is understandable, it sometimes leads to shortcuts that ignore sustainability. Striking a balance between speed and long-term planning is critical.

Q: Can you share any examples of successful sustainable reconstruction efforts from other post-conflict regions?

A: Absolutely. One example that comes to mind is the reconstruction of Sarajevo after the Bosnian War. The city prioritized integrating green spaces into urban planning, using parks and restored wetlands to manage stormwater and improve air quality. These efforts not only addressed environmental concerns but also created recreational spaces that helped bring communities together.

Another example is Rwanda's post-genocide recovery. The government worked with international organizations to introduce sustainable agricultural practices and renewable energy projects. Solar-powered irrigation systems and reforestation programs revitalized the environment while providing livelihoods for displaced populations.

These projects highlight the importance of a holistic approach—combining environmental, economic, and social objectives to create lasting impacts.

Q: What role does international collaboration play in sustainable reconstruction?

A: International collaboration is crucial. No single country or organization has the resources or expertise to address the multifaceted challenges of post-war reconstruction on its own. International partnerships bring together funding, technical expertise, and innovative technologies that can significantly enhance recovery efforts.

For example, in Iraq, international NGOs worked with local governments to rebuild water infrastructure using smart technologies. These systems reduced water wastage and ensured equitable distribution. Similarly, global funding initiatives like the Green Climate Fund can support projects that integrate renewable energy and sustainable practices into rebuilding efforts.

Ukraine's recovery will benefit greatly from such collaborations, particularly in areas like renewable energy integration and eco-friendly urban planning.

Q: What message would you like to share with policymakers and stakeholders involved in post-war reconstruction?

A: My message is simple: think beyond immediate recovery. The choices made today will shape the future of communities for decades. Sustainable reconstruction is not just an ethical responsibility—it's a practical necessity. By investing in resilience, resource efficiency, and inclusivity, we can ensure that post-war recovery lays the foundation for a healthier, more stable, and prosperous future.

Policymakers should prioritize sustainability in funding decisions, international organizations must foster collaboration, and local communities should be empowered to take an active role in the rebuilding process. Together, we can transform destruction into an opportunity for positive change.

Dr. Maria Novak's expertise underscores the transformative potential of sustainable reconstruction. Her insights remind us that recovery efforts can—and should—go beyond restoration, creating resilient and inclusive systems that support both people and the planet.

5.4 Role of International Organizations in Environmental Rehabilitation

The environmental challenges posed by the Russia-Ukraine war are too vast for Ukraine to address alone. International organizations play a critical role in supporting the country's ecological recovery, bringing much-needed funding, expertise, and collaborative frameworks to the table. From assisting in the restoration of ecosystems to promoting cross-border collaborations, global institutions offer Ukraine both the resources and strategies required to rebuild its environmental resilience. This section explores how international organizations contribute to ecological recovery efforts, the importance of funding and expertise, and the significance of cross-border cooperation in restoring Ukraine's damaged ecosystems.

Contributions of Global Institutions to Ecological Recovery Efforts

International organizations bring decades of experience in post-conflict environmental restoration to support Ukraine. Their contributions span several critical areas:

Restoration of Ecosystems: Global institutions like the United Nations Development Programme (UNDP) and the International Union for Conservation of Nature (IUCN) specialize in ecosystem restoration. They provide technical expertise and logistical support for reforestation projects, wetland rehabilitation, and the reintroduction of displaced wildlife. These organizations also facilitate the use of nature-based solutions to address environmental damage, such as restoring floodplains to reduce flood risks and rehabilitating degraded land to enhance biodiversity.

Demining and Contaminant Cleanup: Organizations such as the Halo Trust and the Geneva International Centre for Humanitarian Demining (GICHD) focus on clearing landmines and unexploded ordnance (UXO) that pose long-term risks to both human and environmental health. By making contaminated areas safe, these groups enable the restoration of agricultural lands, forests, and water systems.

Monitoring and Data Collection: Institutions like the World Wildlife Fund (WWF) and the European Environment Agency (EEA) provide tools and expertise for monitoring environmental recovery. Satellite imagery, GIS mapping, and biodiversity surveys help track progress and identify areas requiring intervention. This data is essential for prioritizing efforts and measuring the success of restoration initiatives.

Capacity Building and Training: Many international organizations invest in building local capacity by training Ukrainian conservationists, farmers, and officials in sustainable land management, ecological restoration, and climate-resilient practices. This empowers local communities to take ownership of long-term recovery efforts and ensures that international support translates into sustainable outcomes.

The Importance of International Funding and Expertise in Ukraine's Recovery

Ecological recovery on the scale required in Ukraine demands significant financial resources and specialized expertise. International funding and knowledge-sharing are crucial for addressing the extensive environmental damage caused by the war.

Financial Support for Restoration Projects: Organizations such as the World Bank, Green Climate Fund, and the Global Environment Facility (GEF) provide grants and low-interest loans for large-scale environmental recovery projects. These funds enable Ukraine to undertake critical initiatives, such as reforestation, soil decontamination, and wetland restoration, without diverting resources from other urgent post-war needs.

Investment in Renewable Energy and Green Infrastructure: International funding bodies can help Ukraine rebuild its energy systems with a focus on sustainability. Investments in renewable energy projects, such as wind and solar farms, can reduce the country's dependence on fossil fuels while supporting its economic recovery. Additionally, funding for green infrastructure, including eco-friendly buildings and public transportation systems, ensures that reconstruction efforts align with global climate goals.

Access to Expertise and Best Practices: Post-conflict ecological recovery requires specialized knowledge that Ukraine may lack due to the scale and complexity of the challenges it faces. International organizations bring a wealth of expertise in areas such as demining, habitat restoration, and water resource management. For instance, the IUCN's experience in rebuilding ecosystems in conflict zones can provide valuable guidance for Ukraine's efforts to restore its forests and wetlands.

Advocacy and Awareness: International organizations play a critical role in advocating for Ukraine's environmental recovery on the global stage. By highlighting the ecological impact of the war, they help mobilize additional resources and encourage other nations to support Ukraine's efforts. This advocacy ensures that environmental issues are integrated into broader discussions of post-war reconstruction and international aid.

Promoting Cross-Border Collaborations for Ecosystem Restoration

Ecosystems do not recognize national borders, making cross-border collaboration essential for effective restoration. Ukraine shares several transboundary ecosystems with neighboring countries, such as the Carpathian Mountains and the Danube River Basin. International organizations facilitate cooperation between Ukraine and its neighbors to ensure coordinated efforts in restoring these critical habitats.

Restoring Transboundary Ecosystems: Initiatives like the Carpathian Convention bring together Ukraine, Poland, Romania, Slovakia, and Hungary to protect and restore the Carpathian Mountains. These collaborations focus on reforestation, wildlife conservation, and sustainable tourism, benefiting both biodiversity and local communities. Similarly, the International Commission for the Protection of the Danube River (ICPDR) works to address pollution and habitat degradation in the Danube Basin, ensuring that Ukraine's efforts align with those of its neighbors.

Facilitating Wildlife Migration: International organizations help establish wildlife corridors across borders, allowing displaced animals to safely migrate and repopulate their natural habitats. These corridors are vital for maintaining genetic diversity and ensuring the long-term survival of species affected by the war.

Coordinated Water Resource Management: The restoration of shared water systems, such as the Dnipro River, requires collaboration between Ukraine and its neighbors. International organizations provide frameworks for joint water resource management, addressing issues such as pollution, sedimentation, and ecosystem health.

Shared Knowledge and Resources: Cross-border collaborations allow Ukraine and its neighbors to share knowledge, expertise, and resources. For example, countries that have successfully restored ecosystems after conflicts, such as Bosnia and Herzegovina, can offer valuable lessons for Ukraine's recovery efforts.

Relevance to Ukraine's Environmental Rehabilitation

International organizations and cross-border collaborations are indispensable to Ukraine's environmental recovery. Their contributions extend beyond funding and expertise, providing a platform for coordinated, large-scale efforts that address the interconnected nature of ecosystems. By leveraging international support, Ukraine can restore its damaged environments more effectively and align its recovery with global sustainability goals.

The involvement of global institutions ensures that Ukraine's recovery is not just about rebuilding but about creating a foundation for a sustainable, resilient future. From the Carpathian Mountains to the Danube Basin, these efforts have the potential to transform war-torn landscapes into thriving ecosystems, benefiting both Ukraine and the wider world.

5.5 Environmental Treaties and Their Role in Post-Conflict Recovery

Environmental treaties and global agreements play a pivotal role in guiding post-conflict recovery efforts, offering frameworks for ecological restoration and sustainable development. For Ukraine, these treaties provide a pathway to align its rebuilding efforts with international environmental commitments while addressing the extensive ecological damage caused by war. Leveraging these global agreements can help Ukraine restore its ecosystems, transition to a sustainable economy, and rebuild in a way that minimizes future environmental harm. However, this alignment also comes with challenges, requiring careful planning and international cooperation.

Leveraging Global Agreements for Ecological Restoration

Environmental treaties serve as blueprints for addressing large-scale ecological challenges and restoring ecosystems in post-conflict scenarios. These agreements offer Ukraine access to international funding, technical expertise, and collaborative networks that can support its recovery.

Frameworks for Restoration Efforts: Treaties like the Convention on Biological Diversity (CBD) provide guidelines for conserving biodiversity, restoring habitats, and promoting sustainable land use. Ukraine can adopt these principles to prioritize the restoration of its damaged ecosystems, such as forests, wetlands, and biodiversity hotspots.

For example, the Ramsar Convention on Wetlands, which Ukraine is a party to, emphasizes the importance of protecting wetlands as critical ecosystems. This treaty can guide the restoration of the Dnipro River's wetlands, which have been severely impacted by pollution and destruction during the conflict.

Access to Financial and Technical Resources: Environmental treaties often include mechanisms for funding and capacity-building. The Global Environment Facility (GEF), which supports initiatives under the CBD and other treaties, can provide Ukraine with financial assistance for large-scale ecological recovery projects. Additionally, treaty frameworks facilitate access to technical expertise and best practices from other nations that have faced similar challenges.

Promoting Accountability and Global Standards: By aligning its recovery efforts with international treaties, Ukraine can demonstrate its commitment to global environmental standards. This not only helps attract international support but also ensures that rebuilding efforts incorporate sustainability and climate resilience from the outset.

The Potential of Treaties Like the Paris Agreement to Guide Recovery Efforts

The Paris Agreement, a landmark treaty focused on combating climate change, has significant relevance for Ukraine's post-war recovery. Rebuilding the country in line with the agreement's goals can help mitigate the long-term environmental impact of the war while positioning Ukraine as a leader in sustainable development.

Reducing Carbon Emissions During Reconstruction: The Paris Agreement encourages nations to adopt low-carbon development strategies. Ukraine can leverage this framework to integrate renewable energy, energy-efficient technologies, and sustainable construction practices into its rebuilding efforts. For instance, using green building materials and renewable energy sources can reduce the carbon footprint of reconstruction projects.

Climate Resilience and Adaptation: The agreement emphasizes the importance of climate adaptation, which is critical for Ukraine given its vulnerability to climate-related risks such as flooding and extreme weather. Post-war recovery provides an opportunity to build climate-resilient infrastructure, restore natural buffers like forests and wetlands, and implement policies that enhance the country's ability to adapt to future challenges.

Access to Climate Finance: The Paris Agreement includes provisions for financial support to help nations transition to sustainable economies. Ukraine can tap into funding mechanisms such as the Green Climate Fund to finance renewable energy projects, ecosystem restoration, and other climate-friendly initiatives.

Promoting International Collaboration: The agreement fosters international collaboration on climate action, enabling Ukraine to work with other nations on joint projects, share knowledge, and benefit from collective expertise. This collaboration is particularly valuable for addressing transboundary environmental challenges, such as restoring the Carpathian Mountains or managing the Danube River Basin.

Challenges and Opportunities in Aligning Ukraine's Rebuilding with International Environmental Commitments

While environmental treaties provide a robust framework for recovery, aligning Ukraine's rebuilding efforts with these commitments presents both challenges and opportunities.

Challenges

Competing Priorities: The immediate need for economic recovery and infrastructure rebuilding may overshadow long-term environmental goals. Balancing urgent reconstruction needs with treaty commitments requires careful planning and political will.

Limited Resources: Ukraine faces significant financial constraints due to the war. Securing funding for large-scale ecological projects may be difficult without sustained international support.

Capacity Gaps: Implementing treaty obligations often requires specialized knowledge and institutional capacity, which may be limited in a post-war context. Building this capacity will take time and resources.

Coordination Across Sectors: Aligning environmental goals with other recovery efforts, such as industrial development and housing reconstruction, requires coordination among multiple stakeholders, including government agencies, local communities, and international partners.

Opportunities

A Green Rebuild: The post-war recovery period offers a unique chance to rebuild Ukraine in a way that prioritizes sustainability. By aligning with treaties like the Paris Agreement and the CBD, Ukraine can adopt innovative practices that reduce environmental harm and support long-term growth.

International Support and Partnerships: Environmental treaties provide a platform for Ukraine to engage with the global community, attract funding, and collaborate on shared challenges. This support can accelerate recovery efforts and ensure that they align with best practices.

Strengthened Environmental Policies: The rebuilding process can serve as a catalyst for strengthening Ukraine's environmental policies and institutions, ensuring better protection of ecosystems and biodiversity in the future.

Economic Benefits of Sustainability: Investing in green technologies and renewable energy can drive economic growth, create jobs, and position Ukraine as a leader in sustainable development.

Relevance to Ukraine's Post-War Recovery

Environmental treaties offer Ukraine a roadmap for integrating sustainability into its recovery efforts, ensuring that rebuilding goes beyond physical infrastructure to include the restoration of ecosystems and natural resources. By leveraging global agreements, Ukraine can align its reconstruction with international climate and biodiversity goals, creating a foundation for long-term resilience and prosperity.

While challenges remain, the opportunities for innovation, collaboration, and leadership in sustainability are immense. By embracing the principles of treaties like the Paris Agreement and the CBD, Ukraine can turn the tragedy of war into a chance to rebuild a greener, more sustainable future that benefits both its people and the planet.

5.6 Integrating Community Participation in Restoration

The success of post-war environmental restoration efforts hinges on the active involvement of local communities. Engaging farmers, conservationists, and everyday citizens in rebuilding ecosystems not only accelerates recovery but also ensures the sustainability of these initiatives over the long term. In Ukraine, where the war has left vast areas of land, water systems, and biodiversity devastated, integrating community participation can foster a sense of ownership, resilience, and shared responsibility. Additionally, eco-tourism and sustainable livelihoods offer promising opportunities to align environmental restoration with economic recovery, creating benefits that extend far beyond ecological health.

Engaging Local Communities in Post-War Environmental Projects

Harnessing Local Knowledge: Communities that live in and around affected areas often have a deep understanding of the local environment, including the flora, fauna, and seasonal patterns. Their knowledge can be invaluable in identifying priorities for restoration, such as which areas need reforestation, where water sources have been contaminated, and which species require protection or reintroduction.

Mobilizing Volunteer Efforts: Local participation in activities such as tree planting, clearing debris, or monitoring wildlife can significantly reduce the costs of restoration projects while building a sense of collective purpose. Schools, universities, and local organizations can mobilize volunteers for community-driven environmental initiatives.

Raising Awareness and Building Support: Community involvement fosters awareness of the importance of environmental health for overall well-being. Public education campaigns can help citizens understand the links between ecosystem restoration, climate resilience, and economic stability, increasing support for restoration projects.

Inclusive Decision-Making: Involving communities in decision-making processes ensures that restoration projects address local needs and priorities. This inclusivity strengthens trust between communities and project stakeholders, making initiatives more effective and sustainable.

Empowering Farmers, Conservationists, and Citizens in the Recovery Process

Supporting Farmers in Sustainable Practices: Farmers, as stewards of much of Ukraine's land, play a critical role in environmental recovery. Post-war restoration efforts can empower them by providing training in sustainable agricultural practices such as crop rotation, organic farming, and soil conservation. These practices not only restore degraded lands but also enhance productivity, ensuring food security for local communities and the nation.

Financial incentives, such as subsidies or grants for adopting green farming techniques, can further motivate farmers to participate in restoration efforts. For example, programs that encourage reforestation of buffer zones around farmlands can prevent soil erosion while improving biodiversity.

Equipping Conservationists with Resources: Conservationists and environmental organizations are on the frontlines of restoration. Providing them with the tools, funding, and training they need enables them to effectively address the ecological damage caused by the war. International partnerships can help Ukrainian conservationists access cutting-edge technologies, such as drones for wildlife monitoring or GIS mapping for habitat restoration.

Engaging Citizens Through Grassroots Initiatives: Everyday citizens have a vital role in rebuilding ecosystems. Grassroots initiatives, such as community gardens, local reforestation drives, or waterway cleanups, can empower citizens to contribute meaningfully. These projects also create opportunities for community bonding and psychological healing after the trauma of war.

Encouraging Youth Participation: Involving young people in restoration projects not only accelerates recovery efforts but also instills a lifelong commitment to environmental stewardship. School programs, internships, and youth-led initiatives can channel the energy and creativity of young Ukrainians into restoring their country's natural heritage.

Benefits of Eco-Tourism and Sustainable Livelihoods in Rebuilding Efforts

Eco-Tourism as a Driver of Economic Recovery: Ukraine's rich biodiversity and natural landscapes, including the Carpathian Mountains and the wetlands of the Dnipro River, have the potential to attract eco-tourists. Post-war restoration efforts can integrate eco-tourism as a key component of economic recovery, generating income for local communities while promoting environmental conservation.

Developing eco-friendly tourism infrastructure, such as hiking trails, wildlife observation platforms, and eco-lodges, creates jobs in rural areas and incentivizes the preservation of natural habitats. Eco-tourism also fosters international interest and support for Ukraine's environmental recovery.

Creating Sustainable Livelihoods: Restoration projects can provide communities with sustainable livelihoods, reducing dependence on industries that harm the environment. For example:

Agroforestry: Combining agriculture with tree planting can enhance soil fertility, increase crop yields, and provide additional income from timber or fruit production.

Renewable Energy Projects: Community-owned solar or wind energy initiatives can generate income while reducing reliance on fossil fuels.

Handicrafts and Eco-Friendly Products: Encouraging thc production of handicrafts or products made from sustainable materials can create income opportunities tied to cultural and environmental preservation.

Linking Restoration to Local Identity and Culture: Many communities have cultural ties to their natural surroundings. Restoration projects that celebrate these connections—for example, by promoting traditional farming methods or preserving culturally significant landscapes—strengthen community engagement and pride.

Building Resilient Economies: By focusing on eco-tourism and sustainable livelihoods, Ukraine can build a more resilient economy that is less vulnerable to external shocks. These sectors provide long-term benefits for communities, including job creation, environmental protection, and improved quality of life.

Relevance to Ukraine's Environmental Recovery

Integrating community participation into environmental restoration offers Ukraine a powerful means of accelerating recovery while fostering social cohesion and economic resilience. By engaging farmers, conservationists, and citizens, Ukraine can ensure that restoration efforts reflect the needs and priorities of local communities. Additionally, eco-tourism and sustainable livelihoods create pathways for aligning economic growth with environmental sustainability, providing lasting benefits for both people and nature.

In the aftermath of war, rebuilding is not just about restoring what was lost but also about creating a better, more sustainable future. Community participation lies at the heart of this vision, enabling Ukrainians to take ownership of their country's recovery and build a legacy of environmental stewardship for generations to come.

Conclusion: Rebuilding or Restoring?

As Ukraine embarks on the challenging journey of post-war recovery, the decisions it makes today will shape the future of its environment, economy, and communities. The question of whether to prioritize rebuilding or restoring must not be seen as a binary choice. Instead, Ukraine has the unique opportunity to integrate restoration and sustainability into its reconstruction efforts, creating a more resilient and ecologically balanced future.

This chapter explored the complexities of Ukraine's environmental recovery, from clearing landmines and decontaminating ecosystems to embracing sustainable reconstruction and leveraging international support. It emphasized the importance of aligning rebuilding efforts with global environmental treaties and engaging local communities as active participants in restoration. By adopting green technologies, fostering cross-border collaborations, and empowering farmers, conservationists, and citizens, Ukraine can turn the devastation of war into an opportunity for renewal.

Sustainable practices and ecological restoration are not just ethical imperatives—they are strategic investments in Ukraine's future. They ensure that the country's natural resources remain a foundation for economic recovery, climate resilience, and biodiversity conservation. By prioritizing restoration over rapid industrialization and aligning its efforts with international commitments, Ukraine can set a global example of how post-war recovery can foster both human and environmental prosperity.

In rebuilding its land, Ukraine is also rebuilding hope—hope for a future where nature thrives alongside its people, and where the scars of war are healed not only on the battlefield but in the forests, rivers, and ecosystems that sustain life. This commitment to restoring both the land and its spirit is the legacy that Ukraine can leave for future generations.

Part II: The Rise of Cyber Warfare

Chapter 6: The Digital Battlefield

The Russia-Ukraine war has not only unfolded on traditional battlefields but also on a rapidly evolving digital front. Cyber warfare, once considered a secondary aspect of modern conflicts, has emerged as a primary weapon of disruption and chaos. The conflict has showcased how digital tools and tactics can be weaponized to target critical infrastructure, destabilize governments, and sow misinformation on an unprecedented scale.

In Ukraine, cyberattacks have targeted power grids, financial systems, and communication networks, exposing vulnerabilities in both national and global systems. These attacks have not only disrupted daily life but have also highlighted the interconnectedness of modern economies and the risks posed to global cybersecurity. This chapter delves into the digital dimensions of the Russia-Ukraine war, examining key cyberattacks, their impact on Ukraine's critical systems, and the broader implications for cyber warfare in the 21st century.

6.1 The Evolution of Cyber Warfare: Lessons from the Russia-Ukraine War

Cyber warfare has transformed the nature of conflict, shifting the battlefield from physical terrains to the boundless domain of the digital world. Unlike traditional warfare, which unfolds with troops and weapons on well-defined battlefields, cyber warfare targets critical infrastructure, information systems, and public trust, often with far-reaching consequences. The Russia-Ukraine war has showcased the evolution of cyber warfare from a supplemental tool to a primary weapon, capable of causing widespread disruption and influencing the trajectory of conflict. This section examines the emergence of cyber warfare as a critical front, the fusion of digital tactics with traditional military strategies, and the valuable lessons this conflict offers for global security.

Cyber Warfare as a New Front in Modern Conflicts

In the digital age, cyber warfare has become an essential component of modern conflicts, operating alongside traditional military strategies while introducing entirely new dimensions of engagement. Cyberattacks uniquely target a nation's critical systems, including power grids, financial networks, and communication infrastructure, often with devastating consequences. These attacks are carried out remotely, offering anonymity and precision while minimizing risks to the attacker.

Cyber warfare is defined by several key characteristics. Unlike physical combat, it does not require troops or weapons on the ground, making it highly cost-effective and accessible to a wide range of actors, from nation-states to hacktivists and criminal organizations. Its disruptive power lies in its ability to cripple essential systems, erode public trust, and create chaos without the need for physical destruction. Moreover, it enables asymmetric warfare, where smaller or less militarily advanced nations exploit vulnerabilities in critical infrastructure to challenge larger powers.

The Russia-Ukraine war has brought these characteristics into sharp focus. From targeting Ukraine's energy infrastructure to disrupting communication networks, cyberattacks have amplified the physical toll of war and challenged the nation's ability to function. This conflict underscores the growing role of the digital battlefield in modern warfare.

The Blending of Traditional Military Strategies with Digital Tactics

The integration of cyber tactics into traditional military operations has redefined warfare, creating a multidimensional battlefield that blurs the lines between physical and virtual engagements. The Russia-Ukraine war illustrates how cyberattacks can complement conventional strategies to achieve broader objectives.

Cyberattacks on critical infrastructure have been used to amplify the effects of physical assaults. For example, disabling a power grid through a cyberattack can intensify the chaos caused by missile strikes on industrial facilities, plunging entire regions into darkness and disrupting civilian life. Similarly, digital tactics have been deployed for psychological warfare. By spreading disinformation through social media and other platforms, cyber campaigns have aimed to undermine trust in government institutions, sow discord among allies, and manipulate public perceptions of the conflict.

Espionage has also been a central component of this hybrid warfare. Ukrainian government agencies, military systems, and private organizations have been frequent targets of cyber intrusions designed to steal sensitive information and disrupt operations. Furthermore, attacks on supply chains and logistics systems have delayed the delivery of essential resources, exacerbating the challenges faced by Ukraine in sustaining its defense and humanitarian efforts.

This fusion of digital and physical strategies has created a new paradigm of conflict, where cyberattacks not only amplify traditional tactics but also open entirely new avenues for disruption and influence.

The Russia-Ukraine War as a Case Study in the Escalation of Cyber Conflict

The Russia-Ukraine war has become a defining case study in the evolution of cyber warfare, showcasing its potential to escalate tensions, disrupt national systems, and influence global geopolitics. The conflict provides several critical lessons for understanding the nature and impact of cyberattacks.

One key lesson is the unprecedented scale and sophistication of cyberattacks during the war. Notable incidents include the 2015 and 2016 attacks on Ukraine's power grid, which caused widespread blackouts and demonstrated the vulnerability of critical infrastructure. These events marked a turning point, illustrating how cyber warfare can extend its impact beyond the digital realm to cause tangible harm.

The war also exemplifies the concept of hybrid warfare, where cyber tactics are seamlessly integrated with military, economic, and political strategies to achieve comprehensive objectives. This approach blurs the distinction between wartime and peacetime, making it increasingly difficult to identify and respond to threats in real time.

The global implications of these localized conflicts highlight the interconnectedness of digital systems. For instance, the NotPetya malware, originally targeting Ukrainian systems, quickly spread worldwide, causing billions of dollars in damage and disrupting businesses across multiple industries. This incident underscores how cyberattacks in one region can have ripple effects far beyond their intended targets.

Attribution and accountability remain significant challenges in cyber warfare. The Russia-Ukraine conflict demonstrates how state-sponsored attacks are often disguised as the work of independent hackers, complicating efforts to hold perpetrators accountable. This lack of transparency makes it difficult to develop effective deterrence strategies, further exacerbating the risks associated with cyber conflicts.

Key Takeaways for Global Cyber Defense

The lessons from the Russia-Ukraine war underscore the urgent need for robust cyber defenses and international collaboration to address the evolving threat landscape. Nations must prioritize securing critical infrastructure, investing in technologies such as intrusion detection systems and AI-powered threat analysis. Frameworks for international cooperation are essential to detect and respond to cyber threats, as the interconnected nature of digital systems requires a coordinated global approach. Additionally, establishing norms and regulations for cyber warfare can help mitigate its impact on global stability, fostering accountability and reducing the risk of escalation.

As the Russia-Ukraine conflict demonstrates, the evolution of cyber warfare demands vigilance, innovation, and collaboration. The digital battlefield is now a defining feature of modern conflicts, and safeguarding these interconnected ecosystems is critical for ensuring resilience in an increasingly volatile world. This new frontier of warfare challenges nations to adapt and cooperate in the face of unprecedented threats, shaping the future of global security.

6.2 Major Cyberattacks: Key Targets and Strategies

The Russia-Ukraine war has highlighted the transformative role of cyberattacks in modern conflicts, revealing how digital tactics can destabilize nations by targeting critical infrastructure, information systems, and governance. These coordinated efforts create a multi-pronged strategy to weaken resilience, disrupt essential services, and erode public trust. By examining the key targets and strategies employed during this conflict, we can better understand the devastating potential of cyber warfare and the urgent need for effective defenses.

Attacks on Critical Infrastructure: Power Grids, Energy Facilities, and Communication Networks

Critical infrastructure lies at the heart of a nation's functioning, making it a primary target in cyber warfare. In Ukraine, cyberattacks have sought to cripple essential systems, amplify physical assaults, and sow widespread disruption.

One of the most notable examples occurred in December 2015, when hackers infiltrated the networks of three Ukrainian energy companies, resulting in power outages that affected 225,000 people. This attack, attributed to Russian state-sponsored groups, demonstrated how cyber tactics could cause tangible harm to civilian life. A follow-up attack in 2016 targeted Kyiv's power grid using Industroyer, a sophisticated malware designed to disrupt industrial control systems. These incidents underscored the vulnerabilities of critical infrastructure to advanced cyber threats.

Ukraine's energy sector has also faced relentless cyberattacks aimed at destabilizing its economy and amplifying the effects of physical strikes. Hackers have targeted oil refineries, pipelines, and storage facilities, disrupting production and compounding the economic toll of the conflict.

Communication networks have similarly been a focal point, as their disruption can cripple coordination among government agencies, military forces, and emergency responders. Early in the conflict, Ukrainian internet service providers were hit with distributed denial-of-service (DDoS) attacks, causing widespread outages and bottlenecks. These attacks not only hampered operations but also contributed to chaos and uncertainty.

Disruption of Media and Information Systems: The Weaponization of Disinformation

Cyberattacks on media and information systems have been weaponized to manipulate public perception, spread propaganda, and destabilize social cohesion. In the Russia-Ukraine war, these tactics have played a pivotal role in the broader strategy of psychological warfare.

Ukrainian media outlets have been frequent targets of cyberattacks designed to undermine their credibility and confuse the public. For instance, hackers infiltrated a major news website to post fake reports claiming government officials had fled the country. Such incidents fueled panic and distrust at a critical juncture in the conflict.

Social media platforms have also been weaponized to amplify false narratives, polarize public opinion, and erode unity within Ukraine and its allies. Bots and fake accounts have disseminated fabricated stories about military successes or atrocities, aiming to weaken morale and create divisions. These campaigns have extended to global audiences, attempting to shift blame for the conflict, undermine international support for Ukraine, and influence allied nations' policies.

The psychological impact of these disinformation campaigns goes beyond spreading falsehoods; they are a form of psychological warfare aimed at lowering resistance and destabilizing public morale. By controlling narratives, attackers shape perceptions and erode trust in institutions, weakening societal resilience in the face of conflict.

Targeting Government Institutions: Undermining Trust and Destabilizing Governance

Government institutions are critical to maintaining order and responding to crises, making them a prime target for cyberattacks. In

Ukraine, these attacks have disrupted governance, compromised sensitive information, and eroded public trust in leadership.

Hackers have infiltrated government networks to steal classified data, disrupt operations, and leak confidential information. Such breaches compromise national security and undermine confidence in the government's ability to protect its citizens. Ransomware attacks have added another layer of disruption, locking essential systems and demanding payments in exchange for access, further diverting resources from critical priorities.

Defacing government websites has been another common tactic. In January 2022, multiple Ukrainian government websites were defaced with threatening messages, warning citizens to "be afraid and expect the worst." These attacks sought to intimidate the population and signal Ukraine's vulnerability to its adversaries.

Election systems have also been targeted, with attempts to manipulate voter databases, spread misinformation about candidates, and create doubts about election outcomes. Such interference erodes trust in democratic processes, contributing to political instability and undermining governance.

Strategic Implications of Cyberattacks

The major cyberattacks during the Russia-Ukraine war reflect a deliberate strategy to exploit vulnerabilities, disrupt essential systems, and manipulate information to achieve both tactical and psychological objectives. These attacks demonstrate the increasing role of cyber warfare in modern conflicts.

The integration of cyberattacks with physical military actions amplifies their impact, creating a more comprehensive and devastating assault on a nation's resilience. The interconnectedness of digital systems also means that cyber conflicts in one region can have global repercussions, as evidenced by the NotPetya malware attack, which originated in Ukraine but caused billions of dollars in damages worldwide.

The difficulty of attributing cyberattacks to specific actors further complicates efforts to hold perpetrators accountable, highlighting the need for international cooperation and improved attribution mechanisms.

Lessons for the Future

The cyberattacks on Ukraine underscore the urgency of strengthening cybersecurity defenses at national and international levels. Enhancing the resilience of critical infrastructure, developing rapid response mechanisms, and fostering global collaboration are essential steps in addressing the evolving threat landscape.

The Russia-Ukraine war has set a precedent for the role of cyber warfare in modern conflicts, demonstrating its capacity to destabilize societies, erode trust, and amplify the impact of traditional military actions. As conflicts increasingly extend into the digital realm, addressing these threats will be paramount to ensuring global stability and security. The lessons from this war serve as a stark reminder of the critical importance of proactive defenses and international cooperation in safeguarding the interconnected world we live in.

6.3 Financial Systems Under Siege

In modern warfare, financial systems have emerged as critical targets for cyberattacks, with far-reaching consequences for economic stability, public trust, and national resilience. The Russia-Ukraine war has highlighted the vulnerability of financial networks, with Ukrainian banks and institutions facing relentless cyberattacks designed to destabilize the economy and amplify the impact of physical conflict. These assaults have also underscored the global interconnectedness of financial systems, exposing vulnerabilities that ripple far beyond Ukraine's borders. This section explores the attacks on Ukraine's financial systems, the strategies employed to create economic disruption, and the broader implications for global financial networks.

Cyberattacks on Ukrainian Banks and the Financial Sector

Cyberattacks on Ukrainian banks have been relentless, targeting their ability to function and undermining public confidence in the country's financial stability. Distributed denial-of-service (DDoS) attacks have been a favored tool, paralyzing online banking, payment processing, and ATM services by overwhelming bank servers with excessive traffic. In February 2022, DDoS attacks disrupted several major Ukrainian banks, including

PrivatBank and Oschadbank, leaving thousands of customers unable to access financial services during a critical period of the conflict.

The theft of sensitive financial data has further compounded these challenges. Hackers have infiltrated bank systems to steal account details, transaction records, and personal information. Such breaches compromise both institutional and individual security, with stolen data often sold on the dark web or used to facilitate identity theft and fraud. These intrusions create long-term vulnerabilities, leaving financial systems and customers exposed to future exploitation.

Beyond technical disruptions, these attacks aim to destabilize public confidence in the financial sector. When customers lose access to their accounts or are unable to withdraw funds, panic ensues, exacerbating the economic strain of war. This erosion of trust has profound implications for Ukraine's financial stability, hindering economic recovery and creating a ripple effect of uncertainty.

Economic Disruption Through Ransomware and Phishing Campaigns

Cyberattacks during the Russia-Ukraine war have extended beyond banks, targeting the broader economy through ransomware and phishing campaigns. These methods create widespread disruption by paralyzing institutions, exploiting individuals, and hindering economic activity.

Ransomware attacks have been particularly devastating, encrypting critical financial data and demanding payments—often in cryptocurrency—for its release. Ukrainian financial institutions, already stretched by the physical conflict, have faced multiple ransomware incidents, forcing them to divert resources toward recovery and security enhancements. The impact extends beyond immediate financial losses, as institutions must also rebuild customer trust and resilience against future threats.

Phishing campaigns have been another prevalent tactic, exploiting the chaos of war to manipulate businesses, government agencies, and citizens. Fraudulent emails and messages disguised as official communications from government bodies or humanitarian organizations trick recipients into revealing sensitive information. In Ukraine, such campaigns have stolen login credentials, spread malware, and siphoned funds, compounding the economic disruption caused by the conflict.

Cyberattacks on supply chains have further destabilized Ukraine's economy, disrupting the movement of goods and inflating costs for both producers and consumers. These disruptions have had cascading effects across key industries, including agriculture and manufacturing, deepening the economic toll of the war.

Global Spillover Effects on Financial Networks

The interconnected nature of global financial systems means that cyberattacks on Ukraine's financial networks have not remained localized. Instead, these incidents have exposed vulnerabilities in international systems, with significant repercussions for businesses and governments worldwide.

Some attacks, initially aimed at Ukraine, have inadvertently spread to global networks. The NotPetya malware attack of 2017, which originated in Ukraine, serves as a stark example. While its primary target was Ukrainian infrastructure, the malware quickly spread across borders, affecting multinational corporations, government agencies, and financial institutions worldwide. The attack caused billions of dollars in damages, underscoring the global risks posed by localized cyber conflicts.

Disruptions to Ukrainian banks have also impacted international payment systems and cross-border transactions, creating delays and uncertainty for global businesses and investors with ties to the region. Sanctions and financial restrictions imposed during the conflict have further complicated these challenges, highlighting the fragility of interconnected financial systems during geopolitical crises.

The attacks on Ukraine have raised awareness among financial institutions globally, prompting many to reassess their cybersecurity defenses. Banks, payment processors, and fintech companies have recognized the need for robust protections against cyber threats, particularly in an era where geopolitical tensions increasingly manifest in the digital realm.

Lessons and Recommendations for Strengthening Financial Cybersecurity

The cyberattacks on Ukraine's financial systems offer critical lessons for safeguarding financial networks in the face of evolving threats. Key recommendations include:

Investing in Cybersecurity Infrastructure: Financial institutions must adopt advanced technologies, such as artificial intelligence-based threat detection, to identify and mitigate attacks in real time. These tools enable faster responses to emerging threats, reducing potential damage.

Collaborative Threat Intelligence Sharing: Governments, financial institutions, and cybersecurity firms should coordinate efforts to share intelligence on emerging threats and develop unified responses to cyber incidents. Collaboration is essential for anticipating and countering complex cyberattacks.

Strengthening Regulatory Frameworks: International organizations like the Financial Stability Board (FSB) and the Bank for International Settlements (BIS) play a crucial role in establishing global standards for financial cybersecurity. These frameworks can promote best practices and encourage compliance across jurisdictions.

Building Public Awareness: Educating businesses and individuals about phishing and ransomware risks can reduce vulnerabilities and foster a culture of cybersecurity awareness. Public participation in safeguarding financial systems is a critical component of resilience.

Relevance to Global Financial Security

The attacks on Ukraine's financial systems underscore the critical need for robust cybersecurity measures, not only within Ukraine but across the global financial sector. The financial system has become a frontline target in cyber warfare, with localized conflicts posing significant risks to the interconnected global economy. Learning from Ukraine's experience, nations and financial institutions must adopt proactive strategies to strengthen defenses, ensure resilience, and mitigate the cascading effects of cyber conflicts.

As the Russia-Ukraine war demonstrates, safeguarding financial networks is not just a national concern but a global imperative. The security of one nation's financial system can influence the stability of many others, making collaborative and comprehensive cybersecurity efforts essential for the future of the global economy.

6.4 The Vulnerability of Power Grids in the Digital Age

Power grids, essential to modern life and economic stability, have become critical targets in the age of cyber warfare. These networks rely on complex digital systems that enhance efficiency and control but also introduce significant vulnerabilities. The Russia-Ukraine war has starkly illustrated the devastating potential of cyberattacks on power grids, where blackouts and cascading failures have amplified the chaos of conflict. This section examines how cyberattacks exploit these vulnerabilities, analyzes the specific challenges faced by Ukraine's energy sector, and explores strategies for safeguarding critical infrastructure against the growing threat of cyber warfare.

How Cyberattacks Exploit Infrastructure Vulnerabilities

Power grids, though integral to national security and daily life, are particularly susceptible to cyberattacks due to their reliance on digital technologies like Supervisory Control and Data Acquisition (SCADA) systems and Industrial Control Systems (ICS). These technologies enable remote monitoring and management of power generation and distribution, but their connectivity also creates exploitable entry points for attackers.

Cyberattacks often begin by exploiting weaknesses in digital systems, such as outdated software, poorly configured networks, or unsecured access points. Attackers may use phishing campaigns or malware to gain initial access, then escalate control over critical components. Supply chain vulnerabilities also provide a backdoor for infiltration; compromising third-party vendors' software can grant malicious actors access to multiple networks simultaneously.

SCADA and ICS are particularly attractive targets because they control the physical processes of power delivery. Disabling or manipulating these systems can cause widespread outages, equipment malfunctions, and cascading failures that disrupt interconnected infrastructure. Compounding these challenges, cyberattacks are often designed to remain undetected for extended periods, allowing attackers to gather intelligence or prepare for coordinated strikes.

The interconnected nature of modern power grids amplifies these risks. Disruptions in one part of the system can propagate rapidly, causing nationwide or even regional blackouts that cripple economies and strain public services.

The Case of Ukraine's Energy Sector: Blackouts and Cascading Failures

Ukraine's energy sector has been a focal point of cyberattacks, highlighting the vulnerabilities of power grids during conflict and demonstrating the potential of cyber warfare to disrupt critical infrastructure on an unprecedented scale.

In December 2015, Ukraine experienced one of the first documented instances of a cyberattack directly targeting a power grid. Hackers used the BlackEnergy malware to infiltrate the networks of three energy companies, remotely disconnecting substations and causing widespread blackouts in the Ivano-Frankivsk region. Approximately 225,000 people were left without power for several hours, showcasing the capacity of cyberattacks to cause immediate and tangible harm.

A year later, in 2016, Ukraine faced another attack on its energy sector. Hackers deployed Industroyer, a sophisticated malware specifically designed to target industrial control systems. This attack caused a temporary blackout in Kyiv, further demonstrating the evolving capabilities of state-sponsored cyber actors. Industroyer's level of sophistication underscored the increasing threat posed to energy infrastructure worldwide.

These attacks went beyond immediate power outages. Cascading failures disrupted interconnected systems, including transportation, communications, and emergency services, amplifying the chaos. The economic toll was significant, as businesses faced operational downtime, supply chains were interrupted, and public services were hindered. Additionally, the attacks undermined public trust in the energy sector's reliability, creating psychological and political pressure on both the government and utility providers.

Strategies for Protecting Critical Infrastructure in a Cyber War

The vulnerabilities exposed by attacks on Ukraine's power grid highlight the urgent need for robust defense mechanisms. Protecting these systems requires a comprehensive approach that incorporates technological, organizational, and policy-driven measures.

Strengthening cybersecurity protocols is a critical first step. Power grid operators must adopt advanced security measures, such as network segmentation, encryption, and multi-factor authentication, to limit attackers' access to critical systems. Regular software updates and patches are essential for addressing known vulnerabilities and reducing the risk of exploitation.

Real-time monitoring and threat detection tools, including intrusion detection systems (IDS) and AI-powered analytics, enhance the ability to identify and respond to cyber threats as they emerge. Security Information and Event Management (SIEM) systems can correlate data from multiple sources, providing operators with situational awareness and the capacity to respond quickly and effectively.

Employee training and awareness are equally vital. Human error remains one of the most common entry points for cyberattacks, making it essential to train personnel to recognize phishing attempts, follow best practices, and respond effectively to incidents. Simulated cyberattack exercises can prepare staff for real-world scenarios, improving organizational readiness.

Developing redundancy and resilience within power grids ensures that critical functions can continue even if part of the system is compromised. Distributed energy resources, such as microgrids and renewable energy systems, can provide backup power during disruptions. Investments in grid resilience measures, such as hardening physical infrastructure and enhancing recovery capabilities, minimize the impact of attacks and accelerate restoration efforts.

Public-private partnerships play a crucial role in securing power grids. Collaboration between government agencies, utility companies, and cybersecurity firms facilitates intelligence sharing, coordinated responses, and the development of innovative security solutions. These partnerships can also help fund and implement large-scale infrastructure upgrades.

On a global scale, establishing international norms and cooperation is essential for addressing cyber threats to power grids. Treaties and agreements that prohibit attacks on critical infrastructure can mitigate risks and hold perpetrators accountable, fostering greater stability in cyberspace.

Relevance to Global Infrastructure Security

The cyberattacks on Ukraine's power grid serve as a stark reminder of the vulnerabilities inherent in modern infrastructure. As power grids become increasingly interconnected and reliant on digital technologies, the risks posed by cyberattacks will continue to grow. Addressing these challenges requires a proactive approach that combines technological innovation, robust defense mechanisms, and international collaboration.

For Ukraine, securing its power grid is essential not only for national security but also for rebuilding its economy and restoring public confidence. The lessons learned from these attacks offer valuable insights for protecting power grids worldwide, ensuring resilience against the evolving threats of cyber warfare. By investing in security and fostering global cooperation, nations can better prepare for the complex challenges posed by this new era of digital conflict.

6.5 The Global Implications of Ukraine's Digital Battlefield

The cyber dimension of the Russia-Ukraine war has illuminated the vulnerabilities inherent in digital ecosystems, serving as a wake-up call for nations, corporations, and institutions worldwide. The unprecedented scale and sophistication of the cyberattacks during this conflict have highlighted the necessity for robust cyber defense strategies, international cooperation, and clearer regulations governing cyber warfare. Ukraine's experience on the digital battlefield has not only reshaped its own approach to cybersecurity but has also influenced global practices and norms. This section explores how Ukraine's experience has impacted global cyber defense strategies, underscored the need for international cooperation, and spurred the development of evolving norms for cyber conflict.

How Ukraine's Experience Has Reshaped Global Cyber Defense Strategies

The cyberattacks faced by Ukraine have revealed critical lessons for governments, businesses, and cybersecurity professionals around the world. These incidents have underscored the vulnerabilities of critical infrastructure, the sophistication of state-sponsored cyber actors, and the interconnectedness of global digital systems.

The attacks on Ukraine's power grids, financial networks, and communication systems have emphasized the importance of prioritizing critical infrastructure protection. In response, many nations have increased investments in cybersecurity measures for essential sectors such as energy, transportation, and finance. Advanced technologies, such as AI-powered threat detection and automated response tools, are being adopted alongside enhanced regulations to ensure that critical systems are fortified against potential threats.

Ukraine's experience has also highlighted the growing importance of offensive cyber capabilities as a deterrent. Many nations are now developing cyber arsenals to retaliate against adversaries or preemptively disrupt malicious operations. However, these developments raise ethical and legal questions, underscoring the need for clear international frameworks to regulate their use.

Additionally, the pervasive use of disinformation campaigns during the conflict has prompted governments to focus on media literacy and public awareness programs. Efforts to counter misinformation have expanded to include simulations and training exercises, helping employees in both the public and private sectors better prepare for cyber incidents. These proactive measures aim to reduce the human vulnerabilities that attackers often exploit.

One of the most significant lessons from Ukraine's digital battlefield is the importance of global intelligence sharing and collaboration. Ukraine's partnerships with international cybersecurity firms and allied governments have demonstrated the value of coordinated responses to cyber threats. This approach has led to stronger alliances, fostering the exchange of intelligence and resources to address emerging risks.

The Role of International Cooperation in Combating Cyber Threats

The transnational nature of cyber warfare means that no single nation can address its challenges in isolation. Ukraine's experience has underscored the necessity of international cooperation in combating cyber threats and strengthening global cybersecurity defenses.

During the conflict, Ukraine received substantial support from allied nations, including technical assistance, intelligence sharing, and resources for incident response. Organizations like NATO's Cooperative Cyber Defence Centre of Excellence (CCDCOE) and bilateral partnerships facilitated these efforts, providing Ukraine with critical capabilities to mitigate and counter cyberattacks. This collective response not only strengthened Ukraine's defenses but also set a precedent for collaborative action against cyber aggression.

Private-sector involvement has been equally pivotal. International cybersecurity firms, such as Microsoft, Cisco, and Palo Alto Networks, played a crucial role in identifying and neutralizing threats targeting Ukraine. These companies provided expertise and tools to address vulnerabilities in real time, highlighting the importance of public-private partnerships in addressing cyber risks. These collaborations have also laid the groundwork for global frameworks for incident response, fostering rapid coordination during large-scale cyberattacks.

Cross-border collaboration has extended beyond Ukraine, prompting nations to establish alliances to ensure that cyberattacks originating in one region do not spread unchecked. Regional organizations like the European Union have strengthened their cybersecurity initiatives, including the EU Cybersecurity Strategy, to enhance collective resilience and cooperation among member states. These efforts aim to create a unified defense against increasingly sophisticated cyber threats.

Impacts on Global Norms and Regulations for Cyber Warfare

The Russia-Ukraine war has intensified the urgency of establishing global norms and regulations for cyber warfare. The absence of clear rules governing state-sponsored cyberattacks has created ambiguity, complicating efforts to attribute attacks and hold perpetrators accountable. This conflict has reignited discussions on defining acceptable behavior in cyberspace and developing mechanisms to enforce compliance.

Attribution and accountability remain among the most significant challenges in cyber warfare. State-sponsored actors often disguise their operations to appear as independent groups, making it difficult to identify the true source of an attack. Ukraine's experience has fueled calls for international mechanisms to investigate and attribute cyberattacks with greater transparency. The establishment of a global cyber tribunal or similar institution could provide a platform for holding perpetrators accountable, serving as both a deterrent and a tool for justice.

The targeting of civilian infrastructure during the conflict, including hospitals, power grids, and water systems, has underscored the need for international agreements to protect these assets. Proposals for a "digital Geneva Convention" have gained momentum, advocating for prohibitions on cyberattacks targeting critical civilian infrastructure. These measures aim to mitigate the human and societal costs of cyber conflicts.

Efforts to develop global norms for cyber warfare have been bolstered by initiatives such as the United Nations Group of Governmental Experts (UNGGE), which works to establish voluntary principles for responsible state behavior in cyberspace. These principles include refraining from attacks on critical infrastructure, cooperating in cyber incident investigations, and avoiding the use of cyberspace to undermine international peace and security.

To encourage compliance, international organizations are exploring incentive-based mechanisms, such as economic benefits, technical support, and public recognition for adhering to cyber norms. This approach seeks to foster accountability and cooperation, promoting a more secure and stable digital environment.

Relevance to Future Cybersecurity

The global implications of Ukraine's digital battlefield extend far beyond the immediate conflict. The lessons learned have reshaped how nations approach cybersecurity, emphasizing the need for stronger defenses, international collaboration, and clear rules for engagement in the digital realm. Ukraine's experience serves as a stark reminder of the interconnectedness of modern digital ecosystems and the far-reaching consequences of cyber warfare.

By adopting these lessons and prioritizing proactive strategies, the global community can work toward a more secure and resilient cyberspace. The challenges highlighted by the Russia-Ukraine war underscore the importance of vigilance, innovation, and cooperation in addressing the evolving threats of cyber warfare. As the digital battlefield continues to expand, collective action will be essential to safeguarding the benefits of connectivity and ensuring a stable future for the global digital economy.

Interview with Ivan Kovalenko, Ethical Hacker and Member of Ukraine's Cyber Defense Network

Ivan Kovalenko is an ethical hacker who has played a pivotal role in Ukraine's cyber defense efforts during the Russia-Ukraine conflict. As part of the grassroots "IT Army of Ukraine," Ivan has collaborated with international cybersecurity experts and local volunteers to combat state-sponsored cyberattacks and safeguard critical infrastructure.

Q: Ivan, thank you for joining us. Can you start by explaining how you became involved in Ukraine's cyber defense efforts?

A: Thank you for having me. When the war began, like many Ukrainians, I felt a deep sense of duty to protect my country. As a cybersecurity professional, I knew my skills could make a difference in defending our digital infrastructure. I joined the "IT Army of Ukraine," a decentralized collective of cybersecurity experts, IT professionals, and volunteers. Our mission was clear: to protect Ukraine's critical systems from state-sponsored attacks and, where possible, disrupt the adversary's cyber operations.

This grassroots movement grew rapidly, with thousands of contributors from within Ukraine and across the globe joining the effort. We operate primarily online, using collaborative platforms to share intelligence, coordinate tasks, and deploy defensive and offensive cyber strategies.

Q: What were the primary goals of the IT Army and similar grassroots cyber defense efforts?

A: Our primary goals have been threefold:

1. **Defending Critical Infrastructure**: Protecting Ukraine's power grids, water systems, financial institutions, and communication networks from cyberattacks was our top priority. These systems are vital for civilian life and national stability.
2. **Countering Misinformation**: A significant part of modern warfare is fought in the information domain. We worked to identify and debunk fake news and propaganda that aimed to destabilize public morale or manipulate global perceptions of the conflict.
3. **Disrupting Adversarial Operations**: Where feasible, we took the fight to the adversary by launching counterattacks on servers, websites, and systems that were part of state-sponsored cyber campaigns. This included taking down propaganda sites and exposing vulnerabilities in their networks.

Q: Can you describe some of the challenges you faced while combating state-sponsored cyberattacks?

A: The challenges were immense. First and foremost, we were up against highly organized and well-funded state-sponsored groups. These adversaries had access to sophisticated tools and techniques, often far beyond what is typically seen in the private sector. Groups like Sandworm and Fancy Bear, linked to Russian state actors, are notorious for their advanced persistent threats (APTs).

Another challenge was coordination. While the IT Army was incredibly effective as a decentralized group, managing such a large and diverse network of contributors came with difficulties. Ensuring that everyone stayed aligned on objectives, maintained operational security, and avoided collateral damage required constant communication and leadership.

Additionally, there was the personal risk. Cyber defense work makes you a target—not just digitally but sometimes physically. We had to remain vigilant about protecting our own identities and ensuring our families and communities were safe.

Q: What role did international collaboration play in your efforts?

A: International collaboration was crucial. Cyber defense is not something that can be done in isolation—especially against a sophisticated adversary. We received invaluable support from global cybersecurity firms, independent ethical hackers, and allied governments.

Firms like Microsoft and Cisco provided tools and expertise to secure our critical infrastructure. For example, Microsoft shared intelligence on emerging threats and helped neutralize malware targeting our systems. Cisco offered hardware and software solutions to strengthen network defenses.

International volunteers were also instrumental. Ethical hackers from around the world joined the IT Army, contributing their skills to identify vulnerabilities, block attacks, and counter adversarial operations. This global solidarity reinforced our defenses and amplified our capabilities.

Q: Can you share any specific success stories from your work in Ukraine's cyber defense?

A: One notable success was during the massive distributed denial-of-service (DDoS) attacks on Ukraine's government websites early in the conflict. These attacks aimed to paralyze public communication and sow chaos. Working together, the IT Army and international partners quickly mitigated the attacks by deploying advanced filtering tools and creating alternative communication platforms for government services.

Another success was in countering misinformation campaigns. One of our teams identified a network of fake social media accounts spreading false narratives about Ukrainian military operations. By reporting and exposing these accounts, we were able to limit their reach and protect public morale.

We also managed to disrupt key adversarial systems. In one instance, we identified vulnerabilities in a propaganda network and temporarily disabled several websites used to spread disinformation. While our work was defensive by nature, these targeted disruptions were critical in reducing the adversary's ability to influence the narrative.

Q: What lessons have you learned from your experiences that could help other nations prepare for cyber warfare?

A: There are several key lessons:

1. **Resilience Is Essential**: Building resilient infrastructure is the first line of defense. This means implementing strong cybersecurity measures, such as firewalls, intrusion detection systems, and regular patching of software, well before a crisis begins.
2. **Decentralized Networks Are Powerful**: The IT Army's decentralized structure allowed us to respond quickly and flexibly to emerging threats. Nations should consider fostering similar networks of cybersecurity volunteers and professionals who can mobilize during emergencies.
3. **Global Collaboration Is Key**: Cyber threats are borderless, and defending against them requires international cooperation. Sharing intelligence, tools, and expertise can make all the difference in combating state-sponsored cyberattacks.
4. **Public Awareness Matters**: Educating citizens about cyber hygiene and misinformation is critical. A well-informed public is less susceptible to phishing attacks and propaganda, reducing the overall effectiveness of cyber campaigns.

Q: Looking forward, what do you see as the biggest challenges in cyber defense?

A: The biggest challenge is the constant evolution of threats. Adversaries are always developing new tactics, tools, and techniques, often leveraging technologies like artificial intelligence and machine learning to create more sophisticated attacks. Staying ahead requires continuous innovation and vigilance.

Another challenge is ensuring that the lessons we've learned are institutionalized. Cyber defense shouldn't be reactive—it needs to be proactive, with governments and organizations investing in cybersecurity as a top priority even during peacetime.

Finally, there's the challenge of maintaining global cooperation. Cybersecurity is a shared responsibility, and as geopolitical tensions rise, ensuring collaboration across borders will be critical for addressing these threats.

Q: What gives you hope as you continue your work in Ukraine's cyber defense?

A: What gives me hope is the resilience and ingenuity of people. Despite the immense challenges, we've shown that even against a well-equipped adversary, a collective effort can make a real difference. The global solidarity we've experienced—from international volunteers to support from private companies—reminds me that when people come together, we can overcome even the most daunting threats.

Most importantly, I believe that the lessons from Ukraine will help the world prepare for future challenges. If we take these lessons to heart, we can build a more secure and resilient digital landscape for everyone.

Ivan Kovalenko's insights offer a compelling look into the frontline of Ukraine's digital resistance. His experiences underscore the importance of grassroots efforts, international collaboration, and proactive preparation in defending against modern cyber threats.

Conclusion: The Digital Battlefield

The Russia-Ukraine war has transformed cyberspace into a central front in modern conflicts, showcasing the immense power and devastating potential of cyber warfare. From crippling critical infrastructure to weaponizing disinformation and targeting financial systems, the conflict has demonstrated how the digital realm can disrupt nations, destabilize societies, and challenge global security.

Ukraine's experience highlights the vulnerabilities inherent in interconnected systems and the growing sophistication of state-sponsored cyber aggression. The attacks on power grids, banks, and communication networks have exposed critical weaknesses that resonate far beyond the conflict zone, emphasizing the need for a proactive and coordinated global response.

This chapter has examined the multifaceted dimensions of cyber warfare during the Russia-Ukraine conflict, focusing on its strategic objectives, key tactics, and global implications. The lessons drawn underscore the urgency of strengthening cybersecurity measures, fostering international cooperation, and establishing clear norms and regulations to govern the conduct of cyber warfare. Protecting critical infrastructure, forming public-private partnerships, and ensuring accountability are vital steps in addressing these complex challenges.

As the digital battlefield becomes a defining feature of modern warfare, vigilance and preparedness are paramount. Ukraine's resilience in the face of relentless cyber aggression serves as both a cautionary example and an inspiring call to action. It demonstrates the importance of collective efforts to secure digital ecosystems and build defenses that can withstand the evolving nature of cyber threats.

The global community must rise to this challenge, investing in innovation, collaboration, and the protection of shared digital resources. By addressing the vulnerabilities exposed by Ukraine's digital battlefield, nations can not only safeguard their own systems but also contribute to a more secure and resilient cyberspace for all.

Chapter 7: Weaponizing Information

In the age of digital connectivity, information has become one of the most powerful weapons in modern warfare. Beyond physical battles and cyberattacks, conflicts are now waged in the minds of individuals and societies. The Russia-Ukraine war has showcased how misinformation, propaganda, and psychological operations (PsyOps) can be weaponized to influence perceptions, sow discord, and destabilize governments. From disinformation campaigns targeting citizens to global efforts to sway international opinion, the manipulation of information has emerged as a critical front in the conflict.

Social media platforms, with their vast reach and algorithm-driven amplification, have played a dual role as both conduits for false narratives and tools for countering misinformation. This chapter examines how the weaponization of information has shaped the Russia-Ukraine war, the role of digital platforms in amplifying or mitigating cyber conflicts, and the broader implications of manipulating public perception on a global scale.

7.1 The Rise of Misinformation and Propaganda in Modern Warfare

In the digital age, warfare extends beyond physical battlegrounds into the realms of information and perception, where misinformation, propaganda, and psychological operations (PsyOps) play pivotal roles. The Russia-Ukraine war exemplifies how these strategies are used to shape narratives, influence public opinion, and disrupt societal cohesion. This section examines the evolution of information warfare, the strategic role of propaganda campaigns, and the impact of PsyOps as tools of modern conflict.

The Evolution of Information Warfare in the Digital Age

Information warfare has evolved dramatically with the advent of digital platforms, revolutionizing how propaganda is disseminated and misinformation is weaponized. Historically, propaganda relied on traditional media like newspapers, radio, and television, targeting specific populations with carefully crafted messages. While effective, these methods were geographically limited and slower to implement.

Today, the internet and social media enable the instantaneous spread of propaganda to global audiences. Algorithms amplify reach, while tailored messaging allows actors to exploit demographic-specific vulnerabilities. This transformation has given rise to misinformation—false or misleading information presented as fact—which spreads faster than efforts to correct it.

The Russia-Ukraine conflict has seen widespread use of misinformation, such as fabricated battlefield reports, doctored images, and false narratives of atrocities. These tactics aim to confuse, manipulate, and mislead audiences, creating chaos and undermining resistance. Hybrid warfare, which integrates information operations with conventional and cyber tactics, has further amplified these effects, blurring the lines between physical and digital engagement. The conflict has showcased this hybrid approach, where cyberattacks on infrastructure complement disinformation campaigns, eroding trust and stability.

Propaganda Campaigns: Shaping Narratives and Influencing Public Opinion

Propaganda campaigns in modern warfare are designed to control the narrative of a conflict, presenting a version of events that aligns with the propagandist's agenda. These campaigns distort facts, omit key details, or fabricate stories to sway both domestic and international audiences.

In the Russia-Ukraine war, both sides have employed propaganda to achieve strategic objectives. Russia has framed its actions as "denazification" and "liberation," seeking to justify its military actions and suppress dissent domestically. Ukraine, on the other hand, has emphasized its sovereignty and victimhood, using these narratives to garner international support.

Social media platforms have become critical channels for propaganda, enabling the rapid dissemination of messages through bots, fake accounts, and coordinated campaigns. Manipulated hashtags, viral content, and trends are employed to amplify messages and create the illusion of widespread support. For instance, Russian-linked accounts have promoted narratives supporting their actions, while Ukraine has used platforms like Twitter and TikTok to appeal for global assistance and solidarity.

Propaganda campaigns often target specific audiences to maximize their impact. Domestically, they aim to boost morale or suppress dissent, while internationally, they seek to influence foreign policy and public opinion. Russia's propaganda has targeted its population to justify the war, while simultaneously attempting to undermine NATO and EU support for Ukraine by exploiting divisions in Western societies.

Psychological Operations (PsyOps) as Tools of Disruption and Control

PsyOps are a powerful component of information warfare, designed to influence emotions, motives, and behavior. In conflict, these operations aim to demoralize adversaries, destabilize societies, and undermine resistance efforts.

During the Russia-Ukraine war, PsyOps have been extensively used to spread fear, confusion, and mistrust. For example, disinformation about imminent attacks, exaggerated casualty figures, or false reports of government collapse were disseminated to erode public confidence and morale. False flag operations, which fabricate evidence or create fake events to blame the opposing side, have also been employed to delegitimize adversaries and justify retaliatory actions.

PsyOps often exploit societal divisions, such as political, ethnic, or cultural fault lines, to amplify discord and weaken unity. In the early stages of the conflict, targeted messages predicted Ukraine's swift collapse, aiming to demoralize both civilians and military personnel. False flag operations accused Ukraine of attacking civilians or engaging in provocations, seeking to shift blame and justify escalations.

The psychological impact of these operations is profound, undermining trust in leadership and institutions. Repeated disinformation campaigns have targeted public confidence in Ukraine's government and Western allies, complicating efforts to maintain resistance and unity.

The Implications of Weaponized Information

The Russia-Ukraine war highlights the strategic importance of information warfare as a tool for manipulation and disruption. Misinformation, propaganda, and PsyOps have proven capable of influencing decisions on the battlefield, in boardrooms, and in international diplomacy. These tactics extend the reach of modern conflicts, where narratives can shape perceptions and outcomes far beyond the physical warzone.

The weaponization of information underscores the need for robust countermeasures. Media literacy programs, fact-checking initiatives, and international agreements for the ethical use of information are essential in addressing these challenges. As the boundaries between truth and falsehood blur, societies must remain vigilant against the exploitation of information as a weapon of war.

The lessons from the Russia-Ukraine conflict serve as a crucial blueprint for understanding and combating the dangers of weaponized information. In a digital age where narratives can shape the course of events, protecting the integrity of information systems is as vital as defending physical infrastructure. The rise of information warfare demands a collective commitment to resilience and truth in the face of deliberate manipulation.

7.2 Social Media: A Double-Edged Sword in Cyber Conflicts

Social media has become an essential battleground in modern warfare, providing a platform for the rapid dissemination of both information and misinformation on a global scale. During the Russia-Ukraine war, social media has played a dual role: amplifying conflict through false narratives and serving as a tool for countering disinformation and rallying support. Its influence extends beyond the immediate region, shaping public opinion, driving international discourse, and even influencing policy decisions. This section examines how social media spreads misinformation and panic, the role of algorithms in perpetuating false narratives, and the efforts of platforms to combat misinformation while promoting transparency.

Amplifying Conflicts: How Social Media Spreads Misinformation and Panic

Social media's speed and reach have revolutionized communication, but these same qualities make it a potent vehicle for misinformation, escalating conflicts and exacerbating uncertainty.

The immediacy of social media allows false information to spread faster than it can be fact-checked. During the Russia-Ukraine war, viral posts have included fabricated battlefield developments, false claims of government actions, and doctored images. For example, false reports of Ukrainian forces retreating en masse or Russian troops achieving decisive victories have manipulated perceptions, eroding morale and fostering confusion.

This rapid dissemination of misinformation often fuels panic among civilian populations. Reports of imminent attacks, infrastructure collapse, or food shortages have driven fear-driven behaviors such as hoarding and mass evacuations. These tactics aim to undermine public confidence in government institutions by portraying them as unprepared or incompetent.

Visual content—images and videos—has been weaponized to amplify the impact of misinformation. Out-of-context or doctored visuals, such as images of atrocities or destruction sourced from unrelated events, are used to provoke emotional reactions and shape public opinion. The rise of deepfakes and AI-generated media has further blurred the line between fact and fabrication, with fabricated videos of leaders or alleged war crimes becoming tools of manipulation.

The Role of Algorithms in the Proliferation of False Narratives

Social media platforms rely on algorithms to manage vast amounts of data, optimize engagement, and curate user experiences. While these systems are essential for platform functionality, they also contribute significantly to the spread of misinformation.

Algorithms create echo chambers by showing users content aligned with their preferences and past interactions. This reinforcement of beliefs, regardless of accuracy, deepens divisions and makes it harder to establish consensus on factual information. During the Russia-Ukraine war, these echo chambers have amplified polarizing narratives, fostering entrenched perspectives.

Engagement-driven algorithms prioritize content that generates likes, shares, and comments, inadvertently amplifying sensational or emotionally charged misinformation. Viral narratives, such as exaggerated casualty figures or fabricated battlefield victories, often gain more traction than accurate but less dramatic reports, skewing public perception.

Coordinated bot networks and fake accounts exploit algorithms to spread misinformation at scale. These entities flood platforms with false narratives, creating the illusion of widespread support or dissent. Both pro-Russian and pro-Ukrainian accounts have leveraged bot networks to manipulate public discourse, showcasing how algorithms can be weaponized in information warfare.

Efforts by Platforms to Combat Misinformation and Promote Transparency

Recognizing their role in the spread of misinformation, social media platforms have introduced measures to counter false narratives and enhance transparency. However, these efforts face significant challenges.

Content moderation and fact-checking initiatives aim to identify and remove false information. Platforms like Facebook, Twitter, and YouTube have partnered with fact-checking organizations to flag misleading posts and provide accurate context. For instance, posts spreading false claims about troop movements or civilian casualties during the Russia-Ukraine conflict have been labeled or removed to prevent further dissemination.

Platforms are also working to improve algorithmic transparency. Features such as Twitter's explanations for why specific content appears in users' feeds or Facebook's misinformation labeling tools aim to reduce the impact of false narratives. These measures promote a more balanced information ecosystem while addressing concerns about algorithmic bias.

Media literacy programs have been launched to empower users to critically evaluate the content they encounter online. In Ukraine, such initiatives have helped citizens navigate the flood of conflicting information during the war, equipping them to identify reliable sources and debunk false claims.

Collaboration with governments and organizations has become a cornerstone of social media platforms' efforts to combat disinformation. For example, Meta (formerly Facebook) has worked closely with Ukrainian authorities to counter disinformation campaigns targeting citizens. These partnerships aim to create coordinated responses to the complex challenges posed by information warfare.

The Dual Nature of Social Media in Conflicts

Social media's influence in the Russia-Ukrainc war highlights its dual nature: it is both a weapon of disinformation and a tool for transparency and resistance. While misinformation has been used to manipulate narratives, social media has also served as a platform for documenting war crimes, sharing real-time updates, and rallying international support for Ukraine. This duality underscores the importance of balancing freedom of expression with accountability and truth.

Lessons for the Future

The role of social media in the Russia-Ukraine war underscores its power to shape conflicts, public opinion, and policy. Moving forward, addressing the challenges posed by misinformation while preserving the benefits of digital connectivity requires collaboration between governments, platforms, and users. Key strategies include:

✓ Developing sophisticated tools for detecting and mitigating false narratives, including advancements in AI and machine learning.

✓ Ensuring algorithms prioritize accuracy and transparency over sensationalism, balancing user engagement with ethical responsibility.

✓ Promoting media literacy and critical thinking to empower users to discern truth from falsehood in the digital landscape.

By addressing these challenges, social media can evolve into a force for truth and accountability, minimizing its misuse in future conflicts. The lessons from the Russia-Ukraine war provide a roadmap for shaping the future of information in an interconnected world, emphasizing the need for vigilance, cooperation, and resilience in the digital age.

Interview with Sofia Petrova, Social Media Analyst and Fact-Checking Advocate

Sofia Petrova is a social media analyst with years of experience in identifying and countering disinformation campaigns. She currently works with a global fact-checking organization, focusing on mitigating the impact of misinformation in conflict zones like Ukraine.

Q: Sofia, thank you for joining us. Can you start by explaining how misinformation has shaped global perceptions of the Russia-Ukraine conflict?

A: Thank you for having me. Misinformation has played a pivotal role in shaping how people perceive the Russia-Ukraine conflict. It's a key tool of modern warfare, designed to manipulate public opinion, erode trust in institutions, and amplify divisions both locally and globally.

In the case of Ukraine, disinformation campaigns have been weaponized to serve multiple purposes. On one hand, there are efforts to undermine Ukrainian resilience by spreading false narratives about military operations, economic conditions, or public morale. On the other hand, these campaigns aim to influence international audiences, framing the conflict in ways that align with the aggressor's geopolitical goals.

For example, early in the conflict, false reports about Ukrainian leadership fleeing the country circulated widely. This narrative sought to create an impression of chaos and instability, which could have weakened international support for Ukraine. These tactics are effective because they exploit emotional triggers and are amplified by the speed and reach of social media.

Q: What makes social media such a powerful platform for disinformation campaigns?

A: Social media is an incredibly potent tool for disinformation because of its immediacy, accessibility, and ability to reach a global audience. Platforms like Facebook, Twitter, and Telegram allow information—both true and false—to spread faster than traditional media ever could.

Disinformation campaigns leverage the algorithms of these platforms, which prioritize content that generates high engagement. This often means sensationalized or emotionally charged posts are more likely to go viral, regardless of their accuracy. False narratives are particularly effective because they exploit cognitive biases—people are more likely to believe information that aligns with their existing beliefs or fears.

Another factor is the anonymity and decentralization of social media. Troll farms, bots, and fake accounts can flood platforms with coordinated messages, creating the illusion of widespread support for false narratives. For example, during the Russia-Ukraine conflict, fake accounts were used to amplify propaganda about the conflict's origins, painting Ukraine as the aggressor and justifying military actions.

Q: What challenges do fact-checking organizations face in combating misinformation during a conflict?

A: Fact-checking during a conflict presents unique challenges. First, the sheer volume of misinformation is overwhelming. Disinformation campaigns are well-coordinated and often involve thousands of posts, videos, and images being circulated simultaneously across multiple platforms. Fact-checkers simply can't keep up with this flood in real time.

Second, there's the challenge of **language barriers and localization**. Misinformation in a conflict zone is often tailored to specific regions and languages, making it harder for global fact-checking organizations to identify and debunk. In Ukraine, disinformation is spread not only in Ukrainian and Russian but also in English and other global languages to target international audiences.

Third, **credibility and trust** are major hurdles. Even when fact-checkers debunk false claims, their corrections may not reach the same audience that consumed the original misinformation. Worse, those who are deeply entrenched in a false narrative may dismiss fact-checking efforts as biased or part of a larger conspiracy.

Finally, the speed of misinformation outpaces the speed of verification. Fact-checking takes time—it requires cross-referencing sources, analyzing evidence, and consulting experts. In contrast, disinformation is designed to spread quickly, making it difficult to contain once it goes viral.

Q: Can you share any specific examples of disinformation from the Russia-Ukraine conflict and how it was addressed?

A: One prominent example was the false claim that Ukraine was operating secret biological weapons laboratories funded by the United States. This narrative was widely circulated on social media platforms, often accompanied by manipulated images or documents that appeared to provide "evidence." The story gained traction globally, creating fear and suspicion among international audiences.

Fact-checking organizations worked quickly to debunk this claim. We traced the origins of the narrative to state-sponsored sources and highlighted inconsistencies in the supposed evidence. Experts from the World Health Organization and independent scientific communities were brought in to refute the allegations.

While we were able to limit the narrative's reach in some circles, it continued to circulate in certain communities. This highlights the resilience of disinformation and the need for proactive measures to prevent its spread in the first place.

Q: What strategies are most effective in countering disinformation campaigns?

A: Combating disinformation requires a multi-faceted approach:

1. **Proactive Education**: Raising awareness about how disinformation

works is one of the most effective strategies. Media literacy programs can teach people how to identify red flags in news stories, verify sources, and question sensationalized claims.

2. **Platform Accountability**: Social media platforms must take responsibility for the content they amplify. This includes refining algorithms to reduce the visibility of misleading posts, investing in automated tools to flag suspicious activity, and partnering with fact-checking organizations to provide context for controversial claims.
3. **Rapid Response Networks**: Fact-checkers and cybersecurity experts need to collaborate to identify and address disinformation as it emerges. Real-time monitoring tools, combined with swift publication of fact-checked material, can help limit the spread of false narratives.
4. **Transparency in Fact-Checking**: Fact-checking organizations must remain transparent about their methods and sources to maintain credibility. Providing clear, evidence-based explanations helps build trust with the public.
5. **International Collaboration**: Disinformation knows no borders. Governments, NGOs, and private organizations must work together to share intelligence, coordinate responses, and strengthen global resilience against propaganda.

Q: Are you optimistic about the ability to combat disinformation effectively?

A: I am cautiously optimistic. While the challenges are significant, the global response to disinformation has been improving. More people are aware of the threat it poses, and organizations are developing better tools and strategies to counter it. Governments and social media platforms are beginning to take the issue more seriously, which is an important step forward.

That said, combating disinformation requires sustained effort. It's not something that can be solved overnight. The key is persistence—building systems and educating communities so that misinformation becomes less effective over time. The resilience of truth depends on our collective commitment to protecting it.

Q: What message would you like to share with readers about the role they can play in countering disinformation?

A: My message is simple: be vigilant. Every individual has a role to play in countering disinformation. Before sharing a post, take a moment to verify its source. Question sensational claims, especially those that evoke strong emotional reactions. And most importantly, engage in constructive dialogue. Disinformation thrives in environments of division and distrust—by fostering understanding and critical thinking, we can weaken its impact.

Sofia Petrova's insights highlight the critical role of fact-checking and public awareness in combating disinformation. Her experiences remind us that while the challenges are daunting, the fight against misinformation is essential for preserving truth and trust in an increasingly polarized world.

7.3 Cyber Manipulation of Global Perceptions About the War

In modern conflicts, battles are not only fought on the ground but also waged in the global court of public opinion. The Russia-Ukraine war exemplifies how cyber manipulation can shape narratives, influence alliances, and sway public sentiment on a global scale. Advanced tools such as deepfakes, AI-generated content, and coordinated influence campaigns have blurred the boundaries between truth and falsehood, demonstrating the growing importance of information as a strategic weapon. This section examines how international audiences are targeted, the role of emerging technologies in narrative shaping, and the broader implications of influence campaigns in the global context.

Targeting International Audiences: Framing the Conflict for Global Impact

Modern conflicts extend beyond physical borders, with competing narratives vying for control of the international discourse. Russia and Ukraine have each sought to frame the war to align with their strategic goals, using targeted messaging to influence diplomatic decisions, public opinion, and geopolitical alliances.

Russia has positioned its actions as a defensive measure against NATO expansion, framing the invasion of Ukraine as a mission to "liberate" the region. This narrative aims to justify its actions domestically and internationally, while reducing condemnation from neutral or aligned nations. Conversely, Ukraine has highlighted its sovereignty and resilience, portraying itself as a victim of aggression to secure military and humanitarian aid from Western nations.

Global media and social platforms have become key arenas for this narrative contest. Russia has relied on state-run outlets like RT and Sputnik to disseminate its messaging, targeting audiences in Africa, Asia, and Latin America to counter Western influence. Ukraine, leveraging platforms such as Twitter and TikTok, has humanized the conflict by sharing real-time updates and personal stories, rallying international solidarity and grassroots support.

These campaigns often polarize public opinion, creating divisions within international communities. For example, debates over sanctions, energy policies, and military aid for Ukraine have been shaped by tailored messaging designed to exploit ideological fault lines and sway opinions.

The Use of Deepfakes and AI-Generated Content in Shaping Narratives

The advent of deepfakes and AI-generated content has added a new layer of complexity to information warfare, enabling actors to fabricate highly convincing but false narratives with unprecedented precision and scale.

Deepfakes have been particularly impactful in the Russia-Ukraine war. For instance, fabricated videos purportedly showing Ukrainian President Volodymyr Zelensky surrendering were circulated to demoralize Ukrainian citizens and erode trust in their leadership. Although quickly debunked, these videos showcased the disruptive potential of deepfake technology.

AI-generated content, including images and text, has been employed to flood social media with fabricated materials that reinforce specific narratives. For example, doctored images of supposed atrocities or battlefield events have been used to provoke emotional responses, sway public opinion, and obscure factual reporting.

The sophistication of these tools presents significant challenges for detection. Even experienced analysts can struggle to identify deepfakes and AI-generated media, while fact-checkers are often overwhelmed by the sheer volume of fabricated content. This highlights the critical need for advanced detection technologies and public education to mitigate the impact of AI-based manipulation.

Influence Campaigns to Sway Global Alliances and Public Sentiment

Influence campaigns during the Russia-Ukraine war have been strategically designed to shape diplomatic alignments, sway public sentiment, and impact policy decisions on a global scale.

Russia's campaigns have targeted Western unity, seeking to exploit divisions within NATO and the European Union by amplifying debates over sanctions and military intervention. In contrast, Ukraine has focused on rallying international support, emphasizing the broader implications of the conflict for democracy and international law.

Specific regions have been targeted to achieve strategic objectives. Russia has concentrated efforts on Africa, Asia, and Latin America, portraying the conflict as a proxy war fueled by Western interests and advocating for alternative alliances. These narratives aim to bolster Russia's partnerships in these regions and mitigate the effects of economic sanctions.

Influence campaigns also mobilize public sentiment, using emotionally charged narratives to drive grassroots support for or against specific actions. For example, Ukraine has highlighted the humanitarian crisis to galvanize international aid efforts, while pro-Russian narratives have attempted to discredit Western intervention by framing it as hypocritical or self-serving.

Cyber tools such as bots, fake accounts, and coordinated networks amplify these campaigns, flooding social media with messages that manipulate trending topics and drown out opposing voices. This creates a contested information environment where narratives compete for dominance and public perceptions are shaped on a global scale.

Implications of Cyber Manipulation on Global Perceptions

The widespread use of cyber manipulation has significant implications for global trust in information. As false narratives proliferate, distinguishing fact from fiction becomes increasingly difficult, eroding public confidence in media, institutions, and governments. This undermines democratic discourse and complicates efforts to build international consensus on critical issues.

Manipulated narratives often have unintended global spillover effects, influencing unrelated domains such as energy markets, migration debates, and geopolitical alliances. For instance, false claims about the conflict's impact on global food supplies or energy prices have shaped international policy discussions, demonstrating how localized conflicts can reverberate worldwide.

The Russia-Ukraine war underscores the urgent need for international cooperation to combat cyber manipulation. Establishing norms and regulations for information warfare, improving media literacy, and holding perpetrators accountable are essential steps in mitigating the impact of influence campaigns.

Relevance to Future Conflicts

The cyber manipulation of global perceptions during the Russia-Ukraine war highlights the growing importance of information as a strategic asset in modern warfare. With advanced technologies and influence campaigns shaping the narrative of conflicts, the ability to control perceptions has become a critical factor in determining outcomes.

As tools for cyber manipulation evolve, nations and organizations must prioritize strategies to address these challenges. Investments in detection technologies, transparency in information sharing, and resilience through public education will be crucial in countering the effects of manipulated narratives. The lessons from this conflict provide valuable insights into managing the complex dynamics of information warfare and preparing for its role in future conflicts.

7.4 The Role of Independent Media and Fact-Checking Organizations

In an era dominated by the rapid spread of misinformation, independent media and fact-checking organizations have become indispensable in ensuring access to credible information. The Russia-Ukraine war has highlighted their critical role in challenging disinformation campaigns, preserving journalistic integrity, and safeguarding the public's right to truth in a conflict-driven information environment. This section examines how these entities counter misinformation, the initiatives they employ to dismantle false narratives, and the challenges they face in conflict zones.

Countering Misinformation: The Importance of Credible Journalism

Amid the chaos of war, misinformation thrives, exploiting uncertainty and the absence of verified information. Independent media outlets serve as crucial pillars of truth, providing accurate reporting to help citizens and global audiences understand the realities of conflict. For instance, Ukrainian outlets like *Kyiv Independent* have gained prominence for their comprehensive coverage, countering disinformation from state-controlled media and unauthorized sources.

Credible journalism restores trust in an era where faith in media has eroded due to the prevalence of fake news and biased reporting. By adhering to rigorous journalistic standards—emphasizing balanced coverage, transparency in sourcing, and a commitment to truth—independent media distinguish themselves from propaganda. This trust empowers audiences to rely on factual reporting rather than manipulated narratives.

Independent media also function as watchdogs, holding power to account. By exposing wrongdoing and providing unbiased coverage, they prevent the misuse of misinformation to justify atrocities, escalate conflicts, or manipulate public opinion. Their work ensures that governments and non-state actors remain answerable for their actions during war.

Fact-Checking Initiatives to Dismantle False Narratives

Fact-checking organizations like *StopFake*, *Bellingcat*, and *FactCheck.org* have emerged as vital players in countering disinformation during the Russia-Ukraine war. Their efforts focus on verifying claims, analyzing the origins of misinformation, and debunking falsehoods in real time.

Real-time fact-checking is facilitated by advanced tools such as reverse image searches, metadata analysis, and geolocation, enabling these organizations to swiftly address emerging false narratives. For example, fact-checkers have exposed coordinated disinformation campaigns by

Russian state-sponsored groups that fabricated stories about Ukrainian leadership and military actions. By publicizing these findings, they not only dismantle individual falsehoods but also shed light on the broader strategies behind disinformation efforts.

Fact-checking organizations also engage in educational campaigns to improve media literacy. By teaching individuals how to critically evaluate sources, identify fake news, and verify information, they empower citizens to resist manipulation. These campaigns are particularly crucial in conflict zones, where misinformation can exacerbate fear, division, and instability.

Challenges Faced by Independent Media in Conflict Zones

Operating in conflict zones presents significant risks and challenges for independent media and fact-checking organizations. Journalists often face physical dangers, including shelling, abductions, and attacks. In the Russia-Ukraine war, dozens of journalists have been injured, detained, or killed while covering the conflict, underscoring the perilous nature of their work.

Resource limitations further strain these organizations. Many independent media outlets and fact-checking groups operate with minimal funding and small teams, making it difficult to sustain operations during prolonged conflicts. The financial and technological disparity between disinformation campaigns and efforts to counter them often places these organizations at a disadvantage.

Censorship and intimidation add another layer of complexity. Governments and powerful actors frequently target independent media with legal restrictions, online harassment, or direct threats to suppress dissenting voices. In the Russia-Ukraine war, both Russian and Ukrainian journalists have faced pressures that compromise their safety and independence.

Access to reliable information is another persistent challenge. Journalists and fact-checkers often encounter restricted access to key areas, unverified claims from conflicting sources, and difficulties in corroborating information. These limitations can delay accurate reporting, allowing misinformation to spread unchecked.

The Critical Role of Collaboration

Collaboration between independent media outlets and fact-checking organizations enhances their ability to counter disinformation effectively. On-ground reporting by journalists provides critical context, while fact-checkers offer tools and expertise to verify claims. Partnerships between Ukrainian media and global fact-checking networks during the war have strengthened the accuracy and credibility of coverage.

Support from international organizations like UNESCO and the Committee to Protect Journalists (CPJ) is also essential. These groups provide funding, training, and advocacy, enabling journalists and fact-checkers to operate safely and effectively. Efforts to establish international standards for combatting disinformation and protecting media freedom further bolster the work of these entities.

Implications for Democracy and Stability

The role of independent media and fact-checking organizations extends beyond countering disinformation. Their work is essential for preserving democracy, fostering informed decision-making, and maintaining societal stability. By providing credible information and challenging false narratives, these entities empower citizens to resist manipulation and hold power accountable.

Relevance to Future Conflicts

The Russia-Ukraine war has underscored the critical need for robust and well-supported independent media and fact-checking organizations in the digital age. As information warfare evolves, the ability to discern truth from falsehood will become increasingly vital for global peace and security. Strengthening these institutions and addressing the challenges they face—through funding, collaboration, and public education—will be essential to ensuring that reliable information prevails over misinformation in future conflicts.

By upholding the principles of truth, transparency, and accountability, independent media and fact-checking organizations serve as guardians of democracy and stability, offering a critical defense against the weaponization of information in the digital age.

7.5 The Ethical Dilemma of Weaponizing Information

In the modern age of digital connectivity, information has emerged as a potent weapon in conflicts, capable of shaping narratives, influencing public opinion, and manipulating perceptions on a global scale. While strategic communication can rally support and bolster morale, its misuse through disinformation raises profound ethical concerns. The Russia-Ukraine war exemplifies the complexities of weaponizing information, where the lines between advocacy and manipulation blur, leaving long-term impacts on trust, democracy, and global stability. This section explores the challenges of balancing communication and ethics, examines the societal consequences of information warfare, and highlights the need for global norms and standards.

The Fine Line Between Strategic Communication and Manipulation

Strategic communication is a legitimate tool used by governments and organizations to share accurate, persuasive information that aligns with specific objectives. During wartime, this often involves rallying domestic support, maintaining morale, or seeking international alliances. For example, Ukraine has effectively employed social media to highlight the human cost of the conflict and garner global solidarity. When rooted in factual reporting and transparency, such efforts are widely regarded as ethical and necessary.

However, the ethical dilemma arises when strategic communication crosses into manipulation. The dissemination of false or misleading information, such as fabricated narratives, deepfakes, or doctored images, can demonize adversaries, provoke fear, or shift blame. In the Russia-Ukraine conflict, both sides have been accused of employing manipulative tactics, such as exaggerating victories or atrocities, to gain psychological or political advantages.

The distinction often lies in the intent and consequences of these campaigns. Strategic communication aims to inform and persuade, while manipulation seeks to deceive and exploit. However, even emotionally charged content based on truth can raise ethical concerns if it exploits audiences or distorts perceptions, leaving lasting effects on societal trust and understanding.

Long-Term Societal Impacts of Information Warfare on Trust and Democracy

The widespread use of disinformation in conflicts erodes trust in information sources, including governments, media, and institutions. When audiences become skeptical of all information, even accurate reporting, the foundation of informed decision-making weakens. In the Russia-Ukraine war, conflicting narratives and the volume of misinformation have fostered confusion and mistrust, complicating efforts to establish a shared understanding of events.

Information warfare also deepens societal polarization, exploiting existing divisions along ethnic, political, or cultural lines. This "us versus them" mentality destabilizes nations during conflicts and leaves lasting scars on social cohesion and democratic discourse. For example, disinformation campaigns targeting Ukraine's allies have amplified discord within NATO and the European Union, undermining collective unity.

Democratic processes are particularly vulnerable to information warfare. False narratives can influence elections, shape policy decisions, and delegitimize governments. Fabricated claims about leaders' actions or motives can sway voters, while disinformation campaigns erode confidence in democratic institutions. The normalization of these tactics risks entrenching misinformation as an accepted tool of governance, further weakening societal trust and accountability.

Establishing Norms and Ethical Standards for Information Warfare

As information warfare becomes an increasingly central element of modern conflicts, the need for international norms to govern its use is more urgent than ever. These norms must address the ethical boundaries of information campaigns, protect civilian populations from manipulation, and outline consequences for spreading false information.

Existing frameworks, such as the Geneva Conventions, provide a precedent for establishing rules in warfare, and similar principles can be adapted for the information age. For instance, agreements could prohibit the deliberate targeting of civilian populations with disinformation or the use of fabricated content to justify military actions.

Transparency is key to maintaining the integrity of information campaigns. Governments and organizations engaged in strategic communication must clearly distinguish factual reporting from opinion or advocacy, allowing audiences to critically evaluate their messages. Mechanisms for accountability, such as independent oversight bodies or international agreements, can deter the misuse of information warfare and hold perpetrators accountable.

Balancing national security with ethical considerations is another critical challenge. While shaping narratives is an inevitable part of conflict, the deliberate use of false information undermines credibility and risks long-term reputational damage. Training military and governmental personnel in ethical communication practices can help strike this balance, ensuring that strategic objectives are achieved without compromising integrity.

Public education and resilience-building are equally vital. Media literacy programs and critical thinking initiatives can equip citizens to identify and counter misinformation, fostering a more informed and skeptical public. Collaborations with technology platforms, independent media, and civil society organizations are essential to promote accurate information and combat disinformation effectively.

Relevance to the Russia-Ukraine War and Future Conflicts

The ethical dilemmas surrounding information warfare are not unique to the Russia-Ukraine war but reflect a broader trend in modern conflicts. As digital tools and platforms continue to evolve, the potential for both constructive and destructive uses of information will only grow. Addressing these challenges requires a collective effort to establish global norms, foster accountability, and prioritize truth over short-term gains.

The Russia-Ukraine war highlights the need for a nuanced approach to information in conflicts. While strategic communication can be a powerful tool for rallying support, its misuse through disinformation has far-reaching consequences for trust, democracy, and stability. By establishing clear ethical standards, promoting transparency, and investing in media literacy, the global community can ensure that information remains a force for good, even in the most challenging circumstances.

In conclusion, the weaponization of information is both a challenge and an opportunity for modern societies. The lessons from this conflict serve as a call to action for governments, organizations, and individuals to safeguard the integrity of information systems, fostering a future where truth prevails over manipulation in the digital age.

Conclusion: Weaponizing Information

The Russia-Ukraine war has showcased the transformative power of information warfare in modern conflicts. Through the strategic use of misinformation, propaganda, and psychological operations, both state and non-state actors have weaponized information to shape perceptions, manipulate narratives, and destabilize societies. Social media platforms, central to this battlefield, have acted as both amplifiers of falsehoods and tools for countering disinformation.

This chapter has examined the multifaceted nature of information warfare, from the deployment of advanced technologies like deepfakes and AI-generated content to the vital role of independent media and fact-checking organizations in preserving truth. It also delved into the ethical dilemmas of weaponizing information, highlighting the blurred line between legitimate strategic communication and deceptive manipulation.

The long-term consequences of information warfare extend beyond the immediate conflict. The erosion of trust in institutions, the deepening of societal polarization, and the normalization of disinformation threaten democratic processes and global stability. These challenges call for a collective response: the establishment of international norms to govern information warfare, widespread promotion of media literacy, and building societal resilience to manipulation.

As information continues to influence the outcomes of conflicts and shape global perceptions, the lessons from the Russia-Ukraine war underscore the critical need for vigilance, ethical standards, and collaboration. This is not merely a battle for narrative control but a fight to uphold the integrity of information in the digital age. By prioritizing truth and accountability, the global community can work to ensure that information serves as a force for stability and progress rather than division and deception.

Chapter 8: Cyberattacks Beyond Borders

In the interconnected digital age, the consequences of cyber warfare rarely remain confined to the nations directly involved in a conflict. The Russia-Ukraine war has demonstrated how cyberattacks, initially targeting a specific country, can spill over to impact businesses, governments, and individuals worldwide. From crippling malware that disrupts global supply chains to attacks on international financial systems, the ripple effects of cyber warfare highlight the vulnerabilities of an interconnected world.

This chapter delves into the cross-border implications of cyberattacks, examining how localized conflicts can escalate into global crises. It explores the impact on multinational organizations, threats to international security, and the urgent need for collective resilience and global frameworks to manage cyber conflicts. As cyber warfare continues to evolve, the risks to global stability and cooperation have never been more pressing.

8.1 The Spillover Effects of Cyber Warfare

Cyber warfare, unlike traditional military conflicts, often transcends geographic boundaries. What begins as a localized cyberattack can quickly escalate, disrupting systems and services far removed from the initial target. The interconnectedness of global networks amplifies the reach and impact of these attacks, creating ripple effects that can destabilize economies, interrupt services, and compromise security worldwide. This section examines how localized conflicts lead to global cyber disruptions, offers examples of attacks with unintended international consequences, and highlights the vulnerabilities inherent in an interconnected digital world.

Localized Conflicts with Global Consequences

When cyberattacks are deployed in regional conflicts, the intent is often to disrupt the target nation's critical systems, infrastructure, or economy. However, the complexity of digital ecosystems means that these attacks frequently exceed their intended scope, impacting entities in other countries. This phenomenon is particularly evident in sectors reliant on shared technologies, global supply chains, or international networks.

For example, a ransomware attack targeting a single company in Ukraine may inadvertently spread to its international partners or suppliers, triggering outages or data breaches in entirely different regions. Similarly, malware embedded in a localized system can exploit shared vulnerabilities, cascading into a broader crisis that affects businesses, governments, and individuals across the globe.

Notable Cases of Cyberattacks Spreading Beyond Borders

The Russia-Ukraine conflict has provided numerous examples of cyberattacks that have had unintended international repercussions, demonstrating how localized warfare can escalate into global challenges.

The NotPetya Malware (2017): Originally intended to disrupt Ukrainian systems, the NotPetya malware quickly spiraled out of control, spreading globally within hours. This highly destructive malware, disguised as ransomware, crippled multinational corporations, including shipping giant Maersk and pharmaceutical company Merck, resulting in billions of dollars in damages. NotPetya's rapid propagation was a stark reminder of the interconnected nature of modern digital infrastructure and the risks posed by poorly contained cyberattacks.

Attacks on Financial Systems: Cyberattacks targeting Ukraine's financial institutions have occasionally disrupted international financial networks. For instance, coordinated phishing and malware campaigns against Ukrainian banks have inadvertently affected global payment systems and cross-border transactions, creating uncertainty in financial markets.

These incidents underscore how vulnerabilities in one nation's systems can ripple through the global financial ecosystem.

Energy Sector Incidents: Ukraine's energy infrastructure has been a frequent target of cyberattacks, with some incidents affecting neighboring countries reliant on shared energy grids or interconnected systems. For example, power outages caused by malware targeting Ukrainian utilities have had cascading effects on regional energy markets, raising concerns about the fragility of critical infrastructure in an interconnected world.

The Digital Web of Interconnected Vulnerabilities

The root cause of these spillover effects lies in the deeply interconnected nature of modern digital systems. Global supply chains, cloud computing platforms, and shared communication networks create a web of interdependencies that make it difficult to isolate cyber incidents. While this interconnectedness drives efficiency and innovation, it also creates a single point of failure that can be exploited by attackers.

Shared Technologies and Platforms: Many organizations rely on common software, hardware, and cloud providers. A vulnerability in a widely used platform can become a global risk, as attackers can exploit it to infiltrate multiple systems simultaneously. This was evident in the NotPetya attack, which leveraged a vulnerability in a popular accounting software used in Ukraine but had global implications.

Supply Chain Dependencies: Cyberattacks on suppliers or service providers can cascade through the supply chain, disrupting operations for companies and consumers worldwide. This has been a recurring theme in the Russia-Ukraine war, where attacks on logistics and transportation companies in Ukraine have caused delays and losses for global businesses.

Cross-Border Data Flow: The seamless flow of data across borders makes it challenging to contain cyber incidents. Malware or ransomware introduced in one country can easily travel through international networks, affecting companies, governments, and individuals in other nations.

Implications for Global Security

The spillover effects of cyber warfare demonstrate that no nation is immune to the consequences of localized conflicts in the digital age. As attacks transcend borders, they reveal vulnerabilities in global systems that require collective action to address. The Russia-Ukraine conflict serves as a case study in understanding the interconnected risks of cyber warfare and underscores the need for coordinated strategies to prevent and mitigate cross-border cyber threats.

Building resilience against these spillover effects involves improving cybersecurity frameworks, fostering international cooperation, and prioritizing the security of critical infrastructure. Without these measures, the unintended consequences of cyberattacks will continue to pose significant challenges to global stability and economic security.

8.2 Global Businesses and Organizations Caught in the Crossfire

Cyberattacks originating in localized conflicts, such as the Russia-Ukraine war, often extend beyond their intended targets, drawing multinational corporations and global organizations into the fray. These attacks disrupt operations, erode trust, and result in substantial economic losses. By exploiting the interconnected nature of digital and physical supply chains, cyber warfare can create cascading effects that impact industries worldwide. This section examines the economic consequences for global businesses, the disruptions to supply chains, and case studies that highlight the far-reaching impact of cyberattacks on organizations.

Economic Consequences of Cyberattacks on Multinational Corporations

Direct Financial Losses: Cyberattacks can cause immediate financial damage to multinational corporations through data breaches, operational downtime, and ransom payments. Recovery costs, including cybersecurity upgrades, forensic investigations, and legal fees, further exacerbate these losses.

For example, during the NotPetya attack, several multinational companies reported losses in the hundreds of millions, with recovery efforts stretching over months.

Loss of Business Continuity: Operational disruptions caused by ransomware or malware attacks prevent businesses from fulfilling orders, maintaining production schedules, or delivering services. Prolonged downtime erodes customer trust and loyalty, particularly in industries like logistics, healthcare, and finance.

Global shipping giant Maersk faced a near-total shutdown of its IT systems during the NotPetya attack, resulting in an estimated $300 million loss and significant delays in global shipping.

Damage to Brand Reputation: Cyberattacks erode public trust and damage brand reputation, particularly when customer data is compromised or when companies appear unprepared to handle the fallout. Businesses often face backlash from consumers, investors, and regulators, impacting their long-term profitability.

Regulatory Penalties and Legal Challenges: Organizations affected by cyberattacks may face regulatory fines for failing to protect sensitive data. In regions with stringent data protection laws, such as the European Union (under GDPR), non-compliance can lead to significant penalties.

Legal challenges from affected customers or partners further compound the financial and reputational damage.

Supply Chain Disruptions and Their Cascading Effects

Vulnerabilities in Global Supply Chains: Modern supply chains rely on interconnected networks of suppliers, manufacturers, distributors, and service providers. A cyberattack on any node in this network can have cascading effects, disrupting the flow of goods, services, and information.

For example, attacks on Ukrainian logistics companies during the war disrupted the movement of critical goods, including agricultural products and raw materials, affecting industries worldwide.

Delays and Increased Costs: Cyberattacks often cause delays in production and shipping, leading to missed deadlines, increased costs, and shortages of critical goods. These disruptions are particularly damaging in industries with just-in-time supply chains, where delays can halt entire operations.

During the NotPetya attack, disruptions to global shipping and logistics affected industries ranging from automotive to pharmaceuticals, forcing companies to seek costly alternative routes and suppliers.

Impact on Small and Medium Enterprises (SMEs): SMEs that depend on larger corporations for contracts, supplies, or distribution are often collateral damage in cyberattacks. When a multinational corporation is disrupted, its smaller partners may experience cascading effects, such as delayed payments, canceled contracts, or reduced demand.

The extended impact on SMEs amplifies the overall economic damage, particularly in regions with high interdependence on multinational operations.

Case Studies: The Impact of NotPetya and Other Major Attacks on Global Businesses

NotPetya: A Global Economic Catastrophe: Originally launched as a targeted attack on Ukrainian systems, the NotPetya malware quickly spread worldwide, exploiting vulnerabilities in widely used software.

Key Victims:

✓ *Maersk*: The shipping giant faced a near-total IT outage, disrupting its global operations and resulting in an estimated $300 million in losses. The company had to rebuild its IT infrastructure from scratch, a process that took weeks.

✓ *Merck*: The pharmaceutical company reported losses exceeding $670 million due to production shutdowns and supply chain disruptions caused by the malware.

✓ *FedEx/TNT Express*: The logistics company suffered significant losses when its European operations were paralyzed, with recovery costs reaching $400 million.

NotPetya's indiscriminate spread demonstrated the vulnerabilities of interconnected global networks and the potential for localized cyberattacks to escalate into global crises.

Colonial Pipeline Ransomware Attack (2021): While not directly linked to the Russia-Ukraine conflict, the Colonial Pipeline ransomware attack illustrates the global ramifications of cyberattacks on critical infrastructure. The attack disrupted fuel supply across the U.S. East Coast, leading to panic buying and price spikes.

Relevance to Global Businesses: The attack highlighted vulnerabilities in supply chains reliant on critical infrastructure, prompting multinational corporations to reassess their dependencies and cybersecurity measures.

SolarWinds Supply Chain Attack (2020): The SolarWinds attack, believed to be state-sponsored, targeted a widely used IT management platform, affecting thousands of organizations, including Fortune 500 companies and government agencies.

Global Impact: The attack compromised sensitive data and disrupted operations for months, demonstrating how a single compromised vendor can jeopardize entire industries.

Lessons for Multinational Corporations

The experiences of global businesses caught in the crossfire of cyberattacks underscore the need for proactive measures to mitigate risks and enhance resilience:

Strengthening Cybersecurity Measures: Companies must invest in robust cybersecurity frameworks, including multi-factor authentication, regular vulnerability assessments, and advanced threat detection systems.

Improving Supply Chain Resilience: Businesses should diversify suppliers and establish contingency plans to minimize the impact of disruptions. Enhanced collaboration with partners to ensure shared cybersecurity standards is also essential.

Incident Response Preparedness: Developing comprehensive incident response plans and conducting regular simulations can help organizations respond quickly and effectively to cyberattacks, minimizing downtime and losses.

Global Collaboration and Intelligence Sharing: Multinational corporations should participate in industry-wide and cross-border cybersecurity initiatives to share intelligence, coordinate defenses, and address emerging threats collectively.

The Global Imperative for Cybersecurity

As the interconnectedness of global systems deepens, the risks of cyberattacks impacting multinational businesses and organizations will continue to grow. The Russia-Ukraine war has underscored the urgent need for businesses to view cybersecurity as a fundamental aspect of operational resilience. Protecting against these threats requires not only technological solutions but also a culture of collaboration and preparedness across industries and borders. The lessons learned from past incidents, such as NotPetya, provide valuable insights for safeguarding the global economy against the evolving challenges of cyber warfare.

8.3 Threats to International Security and Stability

The rise of cyber warfare has introduced significant threats to international security and stability. Attacks on critical global infrastructure, the potential for escalation and retaliation, and the role of cyber warfare in intensifying geopolitical tensions have turned the digital realm into a high-stakes battlefield. The Russia-Ukraine war has highlighted the vulnerability of essential systems and the challenges of navigating an interconnected world where the impacts of cyber conflict extend far beyond the initial target. This section explores these threats in detail, examining their implications for global security and stability.

Cyberattacks on Critical Global Infrastructure: Energy, Finance, and Communication

Energy Systems Under Siege: Energy infrastructure is a prime target in cyber warfare due to its critical role in national security and economic stability. Attacks on power grids, oil pipelines, and energy companies can cause widespread disruption, leaving millions without access to electricity or fuel.

The Russia-Ukraine war has seen multiple cyberattacks on Ukraine's energy sector, with malware targeting power grids and oil facilities. These attacks not only disrupted local services but also impacted neighboring countries reliant on shared energy resources.

Global implications arise when energy supply chains are interrupted, as seen in the 2021 Colonial Pipeline attack in the U.S., which, while unrelated to the Russia-Ukraine war, demonstrated how such incidents can ripple through international energy markets, causing price spikes and shortages.

Financial Systems at Risk: Cyberattacks on financial institutions threaten the stability of global markets. Banks, payment processors, and stock exchanges are frequent targets, with hackers aiming to steal funds, disrupt transactions, or manipulate markets.

The interconnected nature of financial networks means that attacks in one region can quickly affect international systems. For example, phishing campaigns and ransomware attacks on Ukrainian banks during the war have disrupted cross-border payment systems, delaying transactions and eroding trust in financial institutions.

Prolonged instability in financial systems can lead to economic uncertainty, reduced investment, and diminished global confidence in market integrity.

Vulnerabilities in Communication Networks: Communication networks are critical for governance, emergency response, and public access to information. Cyberattacks on these systems can sever communication channels, spread disinformation, and hinder coordination during crises.

In the Russia-Ukraine war, distributed denial-of-service (DDoS) attacks have targeted government and corporate communication platforms, disrupting operations and delaying responses to both cyber and physical threats. The reliance on global communication networks makes these vulnerabilities a shared risk.

Risks of Escalation and Retaliation in the Cyber Domain

Difficulty in Attribution: One of the greatest challenges in addressing cyberattacks is identifying the perpetrators with certainty. Hackers often disguise their activities to appear as independent actors, complicating efforts to attribute attacks to specific nations or groups.

Misattribution can lead to unintended escalation, where nations retaliate against the wrong actor, intensifying conflicts and creating broader instability.

Escalating Tit-for-Tat Responses: Cyberattacks often provoke retaliation in kind, creating a cycle of escalating conflict in the digital domain. This tit-for-tat dynamic increases the likelihood of collateral damage and unintended consequences.

For example, a retaliatory cyberattack targeting energy infrastructure could inadvertently disrupt critical systems in neutral countries, drawing more actors into the conflict and increasing geopolitical tensions.

Blurred Lines Between Cyber and Physical Warfare: The integration of cyberattacks with traditional military strategies has blurred the distinction between digital and physical conflicts. A cyberattack on critical infrastructure, such as a power grid, can create conditions that justify a physical military response, raising the stakes and the potential for escalation.

The Russia-Ukraine war has demonstrated how cyberattacks can complement physical operations, creating a multi-dimensional battlefield with heightened risks of miscalculation and overreaction.

The Role of Cyber Warfare in Exacerbating Geopolitical Tensions

Undermining Trust Between Nations: Cyberattacks erode trust between nations, as governments view these incidents as acts of aggression or sabotage. Persistent cyber threats fuel suspicion, making diplomatic negotiations and alliances more challenging to sustain.

During the Russia-Ukraine war, accusations of state-sponsored cyberattacks have deepened divisions between Russia and Western nations, undermining efforts to address the conflict through dialogue.

Weaponizing Global Dependence on Technology: Nations with advanced cyber capabilities can exploit global reliance on technology to exert influence or destabilize adversaries. By targeting critical industries or infrastructure, cyberattacks can create leverage in geopolitical disputes.

The widespread use of Russian software and IT services in global industries has raised concerns about vulnerabilities and the potential for exploitation, leading to calls for stricter cybersecurity measures and diversification of technology supply chains.

Disrupting International Cooperation: Cyberattacks on global systems disrupt the collaborative frameworks needed to address shared challenges, such as climate change, public health, and economic development. Mistrust fueled by cyber warfare hampers the ability of nations to work together, creating a fragmented international order.

The weaponization of digital tools in the Russia-Ukraine war has intensified geopolitical rivalries, with nations prioritizing their cybersecurity agendas over collective action.

Influence on Global Power Dynamics: Cyber warfare is reshaping global power dynamics by providing smaller or less militarily advanced nations with a means to challenge dominant powers. This democratization of conflict in the digital realm increases the number of actors capable of influencing international stability.

The proliferation of cyber capabilities also raises the risk of non-state actors, such as hacktivist groups or cybercriminals, becoming significant players in geopolitical conflicts.

Implications for Global Stability

The threats posed by cyberattacks on critical infrastructure, the potential for escalation, and the exacerbation of geopolitical tensions highlight the urgent need for a coordinated international response. Key priorities include:

Strengthening Global Cybersecurity Frameworks

Nations must collaborate to develop robust cybersecurity frameworks that protect critical infrastructure and reduce vulnerabilities. Initiatives like the Budapest Convention on Cybercrime provide a foundation for international cooperation but require broader participation and enforcement.

Promoting Norms and Agreements for Cyber Warfare

Establishing norms for responsible behavior in cyberspace, such as prohibiting attacks on civilian infrastructure, can reduce the risks of escalation and unintended consequences. International organizations, including the United Nations, play a critical role in facilitating these agreements.

Enhancing Attribution Capabilities

Investments in advanced forensic tools and intelligence-sharing mechanisms can improve the ability to attribute cyberattacks accurately. This reduces the risk of miscalculation and strengthens accountability for perpetrators.

Building Resilience Through Public-Private Partnerships

Collaboration between governments, businesses, and cybersecurity firms is essential for protecting global systems and responding effectively to cyber threats. Shared intelligence, coordinated defenses, and joint exercises can enhance resilience against future attacks.

In conclusion, cyberattacks have become a significant threat to international security and stability, with the potential to disrupt critical infrastructure, provoke escalation, and reshape geopolitical landscapes. The lessons from the Russia-Ukraine war underscore the importance of proactive measures to address these challenges, from strengthening global cybersecurity frameworks to fostering collaboration and trust among nations. In an increasingly interconnected world, ensuring the security and stability of digital systems is a shared responsibility that demands collective action.

8.4 Building Resilience Against Cross-Border Cyber Threats

As cyberattacks increasingly transcend borders, affecting nations, businesses, and individuals worldwide, building resilience against these threats has become a global priority. The Russia-Ukraine war has demonstrated how localized conflicts can disrupt international systems, exposing vulnerabilities in interconnected digital ecosystems. Addressing these challenges requires coordinated efforts among nations, robust cybersecurity frameworks, and proactive measures by corporations and institutions. This section explores the importance of international collaboration and intelligence sharing, the need to strengthen global cybersecurity frameworks, and the role of multinational organizations in enhancing their cyber defenses.

The Importance of International Collaboration and Intelligence Sharing

Cyber Threats Know No Borders: Cyberattacks targeting one nation or organization can quickly spread, impacting systems across borders. The interconnected nature of global networks makes it impossible for any single entity to address these threats alone. International collaboration is essential to mitigate risks and respond effectively.

For example, the NotPetya malware attack, initially aimed at Ukrainian systems, caused widespread disruptions to multinational corporations and organizations worldwide, illustrating the need for collective action.

Sharing Threat Intelligence: Timely sharing of threat intelligence among nations and organizations is crucial for identifying and mitigating cyberattacks before they escalate. Platforms such as the Cyber Threat Alliance (CTA) and the European Union Agency for Cybersecurity (ENISA) facilitate collaboration by enabling real-time exchange of information about emerging threats, vulnerabilities, and attack methods.

During the Russia-Ukraine war, partnerships between Ukrainian authorities, allied governments, and cybersecurity firms have played a critical role in defending against cyberattacks and minimizing their impact.

Coordinated Response to Global Incidents: Collaborative response mechanisms ensure that nations and organizations can act swiftly during major cyber incidents. Joint efforts, such as the development of international Computer Emergency Response Teams (CERTs), provide the infrastructure needed to contain and remediate attacks across borders.

The Global Forum on Cyber Expertise (GFCE) is an example of a multi-stakeholder initiative aimed at building global capacity to respond to cyber threats through knowledge sharing and coordinated action.

Strengthening Global Cybersecurity Frameworks and Agreements

Developing International Norms for Cyber Behavior: Establishing norms for responsible behavior in cyberspace is essential for reducing the risks of cyber conflict and escalation. These norms should address issues such as prohibiting attacks on civilian infrastructure, refraining from cyber espionage against critical systems, and avoiding interference in democratic processes.

The United Nations Group of Governmental Experts (UNGGE) has proposed voluntary norms for responsible state behavior in cyberspace, but broader adoption and enforcement are needed to make these norms effective.

Expanding Participation in Cybersecurity Agreements: Treaties like the Budapest Convention on Cybercrime provide a framework for international cooperation in investigating and prosecuting cybercrimes. Expanding participation in such agreements can enhance global efforts to combat cyber threats.

Regional initiatives, such as the European Union's Cybersecurity Strategy and ASEAN's cooperation on cybersecurity, demonstrate the potential of regional frameworks in addressing specific threats and vulnerabilities.

Enhancing Accountability Mechanisms: Holding actors accountable for cyberattacks is critical for deterring malicious behavior. Improved attribution capabilities, supported by international agreements, can help identify perpetrators and impose consequences, such as sanctions or legal action.

Collaborative efforts, such as the NATO Cooperative Cyber Defence Centre of Excellence (CCDCOE), provide resources for forensic analysis and attribution, strengthening accountability.

Aligning Cybersecurity Standards Across Borders: Harmonizing cybersecurity standards and best practices across nations can reduce inconsistencies and vulnerabilities in global systems. Initiatives like the International Organization for Standardization (ISO) provide guidelines for securing critical infrastructure, supply chains, and digital services.

Investing in Cyber Defense for Multinational Corporations and Institutions

The Role of Private Sector in Cyber Resilience: Multinational corporations and institutions are often on the frontlines of cross-border cyberattacks, making their cybersecurity efforts critical to global resilience. These organizations must adopt comprehensive defense strategies to protect their assets, customers, and partners.

Investment in advanced technologies, such as artificial intelligence (AI) for threat detection and machine learning for predictive analytics, can help organizations stay ahead of emerging threats.

Strengthening Supply Chain Security: Cyberattacks targeting supply chains can disrupt operations and compromise sensitive data. Businesses must collaborate with their suppliers and partners to implement stringent cybersecurity measures, including regular audits, secure communication channels, and incident response protocols.

The SolarWinds attack, which exploited vulnerabilities in a widely used IT platform, highlighted the need for robust supply chain security across industries.

Building Incident Response Capabilities: Developing robust incident response plans and conducting regular simulations can prepare organizations for potential cyberattacks. These plans should include clear protocols for containment, recovery, and communication with stakeholders.

Companies such as Maersk, which faced significant disruptions during the NotPetya attack, have since enhanced their incident response capabilities, demonstrating the value of preparedness.

Collaborating with Governments and Cybersecurity Firms: Public-private partnerships are essential for addressing cross-border cyber threats. Governments, corporations, and cybersecurity firms can share expertise, resources, and intelligence to strengthen defenses collectively.

Organizations like the World Economic Forum's Centre for Cybersecurity bring together stakeholders to tackle global challenges and develop innovative solutions.

Path Forward: Strengthening Global Cyber Resilience

The Russia-Ukraine war has underscored the need for a unified approach to countering cross-border cyber threats. Building resilience requires a multi-faceted strategy that includes:

✓ **International Cooperation:** Nations must prioritize collaboration, intelligence sharing, and joint responses to mitigate the impact of cyberattacks.

✓ **Robust Frameworks:** Strengthening global agreements and developing enforceable norms for cyber behavior can create a more secure digital environment.

✓ **Proactive Investments:** Multinational corporations and institutions must view cybersecurity as a core element of their operations, investing in advanced technologies, incident response, and supply chain resilience.

The interconnected nature of today's digital world means that no nation or organization can address cyber threats in isolation. By fostering collaboration and innovation, the global community can build the resilience needed to navigate the complexities of the cyber domain and safeguard international stability.

8.5 The Need for Global Norms and Regulations

The rise of cross-border cyberattacks has highlighted the urgent need for global norms and regulations to manage cyber conflicts and mitigate their impact on international security and stability. The absence of universally accepted frameworks for cyber warfare has created a gray area where malicious actors operate with relative impunity. Addressing this gap requires overcoming challenges in attributing cyberattacks, developing international agreements to govern behavior in cyberspace, and leveraging the role of global organizations like the United Nations. This section explores these elements and their importance in establishing a secure and cooperative digital environment.

Challenges in Attributing Cross-Border Cyberattacks

The Complexity of Attribution: Cyberattacks are notoriously difficult to attribute to specific actors due to the anonymity and obfuscation techniques employed by attackers. Hackers often use proxies, encrypted communications, and compromised systems to disguise their origins.

State-sponsored groups frequently operate under the guise of independent hackers, further complicating efforts to identify the true perpetrators. For example, many cyberattacks during the Russia-Ukraine war have been attributed to state-linked groups such as Sandworm or Fancy Bear, but definitive proof remains elusive.

The Risk of Misattribution: Misattributing a cyberattack can lead to severe consequences, including unjustified retaliation, escalation of conflicts, and diplomatic fallout. This risk makes nations hesitant to act decisively, even when they suspect a specific actor.

For instance, attributing an attack to a nation without concrete evidence could strain international relations, disrupt trade, or provoke military responses, further destabilizing global security.

The Need for Advanced Forensic Capabilities: Accurately attributing cyberattacks requires advanced forensic tools, skilled analysts, and international cooperation. Nations with limited resources or expertise face significant challenges in identifying perpetrators, leaving them vulnerable to repeated attacks.

Collaborative initiatives, such as the European Union Agency for Cybersecurity (ENISA) and NATO's Cooperative Cyber Defence Centre of Excellence (CCDCOE), aim to enhance attribution capabilities and foster trust among member states.

Developing International Agreements to Manage Cyber Conflicts

Establishing Rules of Engagement in Cyberspace: Just as traditional warfare is governed by international laws, cyber warfare requires rules of engagement to prevent escalation and protect civilian infrastructure. These rules should include prohibitions on targeting healthcare, energy, financial, and communication systems during conflicts.

Efforts to establish such rules are already underway, with the United Nations Group of Governmental Experts (UNGGE) proposing voluntary norms for responsible state behavior in cyberspace.

Creating Mechanisms for Conflict Resolution: International agreements should include mechanisms for resolving disputes and addressing cyber incidents diplomatically. These mechanisms could involve third-party mediation, cyber-specific tribunals, or arbitration processes to de-escalate tensions and hold perpetrators accountable.

The establishment of a global cyber arbitration body could provide a neutral platform for addressing grievances and fostering cooperation.

Harmonizing Legal Frameworks: Divergent national laws on cybercrime and cybersecurity create challenges for international cooperation. Harmonizing these frameworks can facilitate joint investigations, extradition of cybercriminals, and enforcement of cyber norms.

The Budapest Convention on Cybercrime serves as a model for aligning legal standards, but broader participation and updates to address emerging threats are needed.

Incentivizing Compliance: To encourage adherence to international agreements, nations must see tangible benefits, such as enhanced security, economic stability, and access to shared intelligence. Sanctions for violations and rewards for compliance can incentivize participation and cooperation.

The Role of the United Nations and Other Organizations in Establishing Cyber Norms

The United Nations as a Global Convener: The United Nations (UN) has been at the forefront of efforts to develop norms and frameworks for cyber governance. Through initiatives such as the Open-Ended Working Group (OEWG) on Developments in the Field of Information and Telecommunications, the UN provides a platform for member states to negotiate and collaborate on cyber issues.

The UN's focus on inclusivity ensures that both developed and developing nations have a voice in shaping the rules of cyberspace, addressing concerns about equitable access to resources and protection.

Regional and Multilateral Organizations: Regional organizations, such as the European Union (EU), the African Union (AU), and the Association of Southeast Asian Nations (ASEAN), play a complementary role in advancing cybersecurity norms. These groups can address region-specific challenges and foster cooperation among neighboring states.

For example, the EU's General Data Protection Regulation (GDPR) has set global standards for data protection and privacy, influencing cybersecurity practices worldwide.

The Role of Private Sector and Civil Society: Private sector entities, including technology companies and cybersecurity firms, are critical stakeholders in the development of cyber norms. Their expertise in identifying threats, securing systems, and mitigating attacks makes them indispensable partners in global efforts.

Civil society organizations, including academic institutions and advocacy groups, provide valuable perspectives on the ethical implications of cyber governance and advocate for the protection of human rights in digital spaces.

Advancing Transparency and Trust: Building trust among nations and organizations is essential for the success of international agreements. Transparency in cyber operations, such as disclosing vulnerabilities and sharing incident reports, can reduce suspicion and foster collaboration.

Joint cyber exercises and simulations, such as those conducted by NATO, can also enhance trust and preparedness among participating nations.

Path Forward: A Cooperative Vision for Cyber Stability

The establishment of global norms and regulations for cyberspace is critical to ensuring international security and stability in the face of evolving cyber threats. Key priorities include:

Improving Attribution Capabilities: Developing advanced tools and collaborative frameworks to accurately identify perpetrators and reduce the risk of misattribution.

Building Consensus on Norms: Encouraging broad participation in existing initiatives, such as the UNGGE and Budapest Convention, and developing new agreements tailored to emerging challenges.

Strengthening Multilateral Institutions: Supporting the efforts of the UN, regional organizations, and private sector partners to enhance cybersecurity governance.

Promoting Public Awareness: Educating citizens about cyber norms and the importance of responsible behavior in digital spaces to foster a culture of accountability.

In summary, the need for global norms and regulations in cyberspace has never been more pressing. The Russia-Ukraine war has demonstrated the far-reaching consequences of cyber conflicts and the challenges of navigating an unregulated digital domain. By addressing attribution challenges, fostering international cooperation, and establishing clear rules for cyber engagement, the global community can create a safer and more stable digital environment. In an era where cyberspace is both a battlefield and a shared resource, the pursuit of collective security and accountability must remain a priority for nations, organizations, and individuals alike.

Conclusion: Cyberattacks Beyond Borders

The interconnectedness of today's digital world has transformed cyber warfare into a global threat. As demonstrated by the Russia-Ukraine war, cyberattacks originating in localized conflicts often transcend borders, impacting multinational corporations, disrupting supply chains, and threatening critical global infrastructure. These attacks highlight the vulnerabilities of interconnected systems and the urgent need for collective action to strengthen cyber resilience.

This chapter explored the wide-ranging implications of cross-border cyber threats, emphasizing the risks to international security, stability, and economic prosperity. From the economic fallout of attacks like NotPetya to the risks of escalation in the cyber domain, the lessons from these incidents underline the importance of international collaboration, robust cybersecurity frameworks, and proactive investments in defense.

Addressing these challenges requires global norms and regulations that govern behavior in cyberspace, ensure accountability, and foster cooperation among nations. By prioritizing the protection of critical infrastructure, sharing intelligence, and aligning efforts across sectors and borders, the global community can mitigate the risks posed by cyberattacks and safeguard the digital ecosystems that underpin modern life.

The digital battlefield is no longer confined to national boundaries, making cyber resilience a shared responsibility. As cyber threats continue to evolve, the lessons learned from the Russia-Ukraine conflict must guide the development of strategies to ensure a secure, stable, and cooperative digital future for all.

Chapter 9: Cyber Defense and Resilience

As cyberattacks have become a defining feature of modern conflicts, the ability to defend against and recover from digital threats has emerged as a critical aspect of national security. The Russia-Ukraine war has underscored the importance of cyber defense and resilience, with Ukraine leveraging a combination of grassroots efforts, international support, and cutting-edge technology to counter an onslaught of cyber threats. This chapter explores Ukraine's digital resistance, the tools and strategies employed to defend against cyberattacks, and the pivotal role played by cybersecurity firms, ethical hackers, and public-private collaborations in building long-term resilience. Through these efforts, Ukraine has provided valuable lessons on navigating the challenges of the digital battlefield.

9.1 Ukraine's Digital Resistance: A Model of Cyber Defense

In the face of relentless cyberattacks, Ukraine has demonstrated remarkable resilience and adaptability in defending its digital infrastructure. The nation's cyber defense efforts have been a collaborative endeavor, combining the expertise of Ukrainian IT professionals, the contributions of hacktivist groups and cyber volunteers, and international support from allies like NATO. Together, these forces have built a robust model of cyber defense that offers valuable lessons for other nations navigating the challenges of cyber warfare.

Grassroots Cyber Defense Efforts: Contributions of Ukrainian IT Experts

The Role of Local Expertise: Ukraine's thriving IT sector, one of the largest in Eastern Europe, has been instrumental in its cyber defense efforts. Ukrainian IT professionals, cybersecurity experts, and software developers have rallied to protect critical systems and respond to ongoing cyber threats.

These experts have worked to secure government networks, financial systems, and communication platforms, often under intense pressure. Their work includes identifying vulnerabilities, mitigating attacks, and implementing real-time defenses.

Emergency Response Teams: Ukrainian IT experts formed ad hoc Computer Emergency Response Teams (CERTs) to coordinate efforts during the conflict. These teams acted as the first line of defense, responding to cyber incidents, analyzing threats, and restoring compromised systems.

Their grassroots approach allowed for rapid decision-making and deployment of resources, minimizing the impact of attacks on critical infrastructure.

Cybersecurity Innovation Under Duress: The war has spurred innovation in Ukraine's cybersecurity practices. Ukrainian IT experts have developed and deployed custom tools, such as anti-malware software and intrusion detection systems, tailored to address the unique challenges posed by state-sponsored cyberattacks.

Collaborative platforms have also emerged, enabling IT professionals to share intelligence and coordinate defenses in real time.

The Role of Hacktivist Groups and Cyber Volunteers

Hacktivists Joining the Fight: Hacktivist groups have played a significant role in supporting Ukraine's cyber defense efforts. These groups, composed of independent hackers and collectives, have launched counterattacks against Russian targets, disrupting propaganda channels, exposing vulnerabilities, and collecting intelligence.

The "IT Army of Ukraine," a decentralized collective of cyber volunteers, was organized early in the conflict to crowdsource talent and resources for cyber operations. Its members, both domestic and international, have been instrumental in countering disinformation campaigns and launching digital offensives against adversaries.

Disrupting Adversarial Operations: Hacktivist groups and cyber volunteers have focused on disrupting Russian propaganda, exposing state-sponsored disinformation campaigns, and targeting Russian military logistics. Their actions have included taking down websites, leaking sensitive data, and compromising communication channels used by adversaries.

These efforts have not only undermined Russian cyber operations but also boosted Ukrainian morale by demonstrating the power of collective resistance in the digital domain.

Ethical Dilemmas and Risks: While hacktivists and cyber volunteers have contributed significantly to Ukraine's cyber defense, their involvement raises ethical and legal questions. Issues such as accountability, potential escalation, and the unintended consequences of their actions remain subjects of debate.

Despite these challenges, their contributions have highlighted the potential of decentralized, grassroots movements in defending against cyber threats.

International Alliances and NATO's Cyber Support

The Role of International Partnerships: Ukraine's cyber defense efforts have been bolstered by strong international alliances. NATO has been a key partner, providing technical expertise, intelligence sharing, and training to enhance Ukraine's cyber capabilities.

NATO's Cooperative Cyber Defence Centre of Excellence (CCDCOE) has played a pivotal role in supporting Ukraine, offering guidance on mitigating attacks, securing critical infrastructure, and improving incident response.

Collaborative Cyber Exercises: Ukraine has participated in joint cyber defense exercises with NATO and other allied nations, improving its preparedness and fostering collaboration. These exercises simulate cyberattacks on critical infrastructure, allowing participants to develop and test coordinated responses.

Such collaborations have strengthened Ukraine's ability to detect, respond to, and recover from cyber incidents, while also fostering trust and knowledge-sharing among allies.

Support from Global Cybersecurity Firms: International cybersecurity companies, such as Microsoft and Cisco, have provided crucial support to Ukraine. These firms have offered tools, resources, and expertise to help defend against sophisticated cyberattacks.

For example, Microsoft identified and neutralized malware targeting Ukrainian systems, while Cisco supplied hardware and software to strengthen network security.

Aid from Friendly Nations: Governments from allied nations have also contributed to Ukraine's cyber defense. Countries such as the United States, the United Kingdom, and members of the European Union have provided funding, equipment, and technical support to bolster Ukraine's digital resilience.

These partnerships have reinforced Ukraine's position on the digital battlefield, ensuring that it has the resources and expertise needed to counter cyber threats effectively.

A Model for Cyber Defense

Ukraine's digital resistance represents a model of cyber defense that blends grassroots innovation, international collaboration, and the contributions of decentralized networks. By leveraging the expertise of IT professionals, the energy of hacktivist groups, and the support of global allies, Ukraine has created a resilient and adaptive approach to cyber warfare. The lessons from this experience offer valuable insights for nations worldwide as they navigate the evolving challenges of the digital age.

9.2 Tools and Strategies for Defending Against Cyberattacks

The increasing sophistication of cyberattacks necessitates equally advanced tools and strategies for defense. The Russia-Ukraine war has underscored the importance of proactive and reactive measures to protect critical infrastructure, mitigate the impact of attacks, and build long-term resilience. From cutting-edge technologies to fundamental best practices, defending against cyber threats requires a multi-layered approach that combines advanced tools, workforce preparedness, and the integration of artificial intelligence (AI) and machine learning (ML). This section explores these key elements in detail.

Advanced Threat Detection and Incident Response Tools

Intrusion Detection and Prevention Systems (IDPS): IDPS are critical components of cyber defense, designed to monitor network traffic for suspicious activity and prevent unauthorized access. These systems use behavioral analytics and signature-based detection to identify potential threats in real time.

During the Russia-Ukraine war, IDPS played a vital role in detecting malware targeting government and financial systems, enabling swift responses to contain threats before they caused widespread damage.

Endpoint Detection and Response (EDR): EDR tools provide visibility into endpoints—such as computers, servers, and mobile devices—that are often the entry points for attackers. These tools monitor for abnormal behaviors, isolate compromised endpoints, and facilitate forensic investigations.

In Ukraine, EDR solutions helped organizations identify ransomware infections and remove malicious software from their networks, minimizing downtime and data loss.

Threat Intelligence Platforms (TIPs): TIPs aggregate data from multiple sources, providing organizations with actionable insights into emerging threats, vulnerabilities, and attack methods. These platforms enable proactive defense by helping organizations stay ahead of potential attackers.

For example, Ukraine leveraged threat intelligence from global cybersecurity firms and allied governments to anticipate attacks and strengthen its defenses.

Automated Incident Response Tools: Automation is a key strategy in responding to cyberattacks quickly and effectively. Automated tools can isolate affected systems, block malicious IP addresses, and restore services without requiring manual intervention.

These tools were particularly useful during the Russia-Ukraine conflict, where the volume of attacks required rapid responses to prevent cascading failures.

Cyber Hygiene Practices and Workforce Training

Cyber Hygiene Basics: Cyber hygiene refers to the routine practices that individuals and organizations implement to secure their systems and data. These include regular software updates, strong password policies, multi-factor authentication (MFA), and data encryption.

In Ukraine, promoting cyber hygiene among employees and citizens was crucial for preventing phishing attacks, one of the most common methods used by attackers to gain access to systems.

Employee Awareness and Education: Human error remains one of the leading causes of successful cyberattacks. Workforce training programs that focus on recognizing phishing emails, avoiding suspicious links, and following security protocols are essential for reducing vulnerabilities.

During the war, Ukrainian organizations launched awareness campaigns to educate employees about common attack vectors and how to report suspicious activity promptly.

Regular Security Assessments: Conducting regular vulnerability assessments and penetration testing helps organizations identify weaknesses in their systems before attackers can exploit them. These practices are critical for maintaining robust defenses in a rapidly evolving threat landscape.

Ukrainian organizations used these assessments to prioritize security investments and reinforce critical systems against likely attack scenarios.

Incident Response Training and Drills: Preparing employees to respond effectively to cyber incidents reduces the impact of attacks. Incident response training and simulated drills ensure that teams can execute recovery plans, restore services, and communicate effectively during crises.

Ukraine's response teams conducted regular exercises to prepare for ransomware attacks, DDoS campaigns, and other cyber threats.

The Importance of AI and Machine Learning in Proactive Defense

AI-Powered Threat Detection: AI and ML algorithms excel at analyzing vast amounts of data to detect patterns and anomalies that indicate potential threats. These technologies can identify sophisticated attacks that traditional tools may overlook, such as zero-day vulnerabilities and advanced persistent threats (APTs).

During the Russia-Ukraine conflict, AI-powered tools helped identify and block malware embedded in seemingly legitimate files, preventing data breaches and system compromises.

Predictive Analytics: ML algorithms use historical data to predict potential attack scenarios and identify high-risk areas. This allows organizations to allocate resources more effectively and implement targeted defenses.

Predictive analytics enabled Ukrainian cybersecurity teams to anticipate the methods and timing of attacks, enhancing their ability to thwart adversaries.

Automated Response and Remediation: AI-driven tools can respond to threats autonomously, minimizing response times and reducing the workload on human teams. For example, AI systems can isolate infected devices, block malicious traffic, and implement patches without manual intervention.

These capabilities were crucial during the conflict, where rapid responses were necessary to contain the spread of ransomware and other malware.

Enhanced Threat Intelligence: AI and ML improve threat intelligence by analyzing data from diverse sources, identifying trends, and generating actionable insights. This helps organizations stay ahead of attackers and adapt their defenses to new tactics.

Ukrainian organizations benefited from AI-enhanced threat intelligence shared by global cybersecurity firms, enabling them to bolster their defenses against evolving threats.

Building a Comprehensive Defense Strategy

Ukraine's experience in defending against cyberattacks highlights the importance of combining advanced tools, strong cyber hygiene practices, and innovative technologies. A comprehensive defense strategy involves not only investing in cutting-edge solutions but also fostering a culture of cybersecurity awareness and preparedness across all levels of an organization.

By integrating these elements, nations and organizations can build resilience against cyber threats, ensuring that they can detect, respond to, and recover from attacks effectively. The lessons from Ukraine serve as a blueprint for navigating the complexities of modern cyber warfare and safeguarding critical systems in an increasingly hostile digital environment.

9.3 The Role of Cybersecurity Firms in the Conflict

The Russia-Ukraine war has underscored the critical role of global cybersecurity firms in defending against sophisticated cyberattacks. As frontline defenders, these companies bring cutting-edge tools, expertise, and resources to protect critical systems and support national governments and private organizations in mitigating cyber threats. Partnerships between cybersecurity firms and government entities have proven essential in responding to the challenges posed by state-sponsored and criminal cyberattacks. This section explores the contributions of these firms, highlights the importance of public-private collaboration, and examines key success stories that illustrate their impact during the conflict.

Global Cybersecurity Companies as Frontline Defenders

Providing Advanced Cybersecurity Solutions: Leading cybersecurity firms such as Microsoft, Cisco, Palo Alto Networks, and CrowdStrike have been instrumental in protecting Ukrainian and allied systems from cyberattacks. These companies offer advanced tools for threat detection, intrusion prevention, and incident response, which have been crucial in mitigating the impact of cyber threats during the war.

For example, Microsoft provided real-time threat intelligence and tools to neutralize malware targeting Ukrainian networks, ensuring the continuity of critical services.

Neutralizing Emerging Threats: Cybersecurity firms continuously monitor for new malware and vulnerabilities, often identifying and neutralizing threats before they can cause significant damage. Their ability to quickly adapt to evolving attack techniques has been a key factor in preventing widespread disruptions.

During the conflict, firms detected and mitigated attacks involving wiper malware and advanced persistent threats (APTs), which were specifically designed to disable Ukrainian infrastructure.

Global Threat Intelligence Networks: The extensive threat intelligence networks maintained by cybersecurity firms provide invaluable insights into emerging attack trends and methodologies. These networks enable the rapid dissemination of information about threats, helping organizations worldwide bolster their defenses.

Companies like Symantec and FireEye have shared detailed reports on the tactics and tools used by state-sponsored groups during the Russia-Ukraine war, aiding global preparedness.

Partnerships between Private Firms and Government Entities

Strengthening National Cyber Defenses: Collaboration between cybersecurity firms and government agencies has been crucial in fortifying Ukraine's cyber defenses. These partnerships provide governments with access to specialized tools, expertise, and resources that may not be available internally.

The U.S. Cybersecurity and Infrastructure Security Agency (CISA) worked closely with private firms to support Ukraine's cyber resilience, sharing intelligence and providing technical assistance.

Coordinated Incident Response: Public-private partnerships have enabled coordinated responses to major cyber incidents. Cybersecurity firms have worked alongside government agencies to investigate attacks, identify vulnerabilities, and implement countermeasures.

During the Russia-Ukraine war, joint efforts between Ukrainian authorities and private firms were instrumental in restoring compromised systems and mitigating the impact of ransomware campaigns.

Training and Capacity Building: Cybersecurity firms have also played a role in training government and private sector personnel, enhancing their ability to detect and respond to cyber threats. These capacity-building efforts have helped create a more resilient cyber workforce in Ukraine.

For example, Cisco's Networking Academy provided training programs focused on cybersecurity best practices and incident response techniques.

Case Studies: Success Stories from Cybersecurity Firms During the Russia-Ukraine War

Microsoft's Threat Intelligence and Mitigation: Microsoft has been a standout contributor in Ukraine's cyber defense efforts. The company's Threat Intelligence Center identified and neutralized several cyber threats, including wiper malware designed to disable Ukrainian systems.

In one instance, Microsoft detected a wave of cyberattacks targeting Ukrainian critical infrastructure and provided tools to block these attacks, ensuring the continuity of essential services such as energy and healthcare.

Cisco's Secure Network Solutions: Cisco provided secure networking solutions to Ukrainian organizations, helping them withstand DDoS attacks and other forms of cyber aggression. The company also deployed its Talos Incident Response Team to assist in mitigating and recovering from cyber incidents.

Cisco's proactive support was critical in maintaining the operational integrity of Ukrainian networks during the conflict.

CrowdStrike's APT Defense: CrowdStrike played a significant role in defending against advanced persistent threats (APTs) linked to Russian state-sponsored groups. The company's endpoint protection tools and threat intelligence services helped Ukrainian organizations identify and neutralize sophisticated cyberattacks.

By sharing intelligence on APT tactics and techniques, CrowdStrike also contributed to global awareness of the cyber strategies employed during the war.

Palo Alto Networks' Infrastructure Protection: Palo Alto Networks provided advanced firewall and cloud security solutions to protect Ukrainian infrastructure from cyberattacks. These solutions enabled organizations to detect and block malicious traffic, ensuring the integrity of their systems.

The company's Unit 42 threat intelligence division also published detailed analyses of cyber threats related to the conflict, enhancing global understanding of the tactics used by adversaries.

Lessons from Cybersecurity Firms' Contributions

The efforts of cybersecurity firms during the Russia-Ukraine war have highlighted several key takeaways:

- ✓ **The Importance of Collaboration:** Partnerships between private firms and government entities are essential for addressing the scale and complexity of modern cyber threats.
- ✓ **The Value of Threat Intelligence:** Sharing actionable intelligence on emerging threats enhances preparedness and resilience across sectors and borders.
- ✓ **The Need for Proactive Defense:** Investing in advanced cybersecurity solutions and capacity building before a crisis ensures organizations are better equipped to handle cyberattacks.

In summary, Cybersecurity firms have been indispensable allies in Ukraine's fight against cyber aggression. Their tools, expertise, and collaborative efforts have helped mitigate the impact of sophisticated attacks and preserve the integrity of critical systems. As cyber threats continue to evolve, the role of these firms will only grow in importance, offering a model for how public and private sectors can work together to build cyber resilience and defend against the challenges of the digital age.

Interview with Markus Engel, Cybersecurity Specialist at NATO

Markus Engel is a senior cybersecurity specialist working with NATO's Cooperative Cyber Defence Centre of Excellence (CCDCOE). With over 15 years of experience in international cyber defense strategies, Markus has been at the forefront of fostering collaboration to address cross-border cyber threats.

Q: Markus, thank you for speaking with us. Can you begin by explaining NATO's role in strengthening cyber defenses, particularly in light of conflicts like the Russia-Ukraine war?

A: Thank you for having me. NATO's role in cyber defense is grounded in our collective security principle: an attack on one is an attack on all. In today's digital age, this extends to cyberspace. The Russia-Ukraine conflict has demonstrated how cyberattacks can destabilize nations and disrupt global systems, making cybersecurity a critical aspect of modern defense strategies.

NATO supports member and partner nations by providing resources, intelligence, and expertise to bolster their cyber defenses. For instance, our Cooperative Cyber Defence Centre of Excellence (CCDCOE) conducts research, training, and simulations to prepare nations for emerging threats. We also facilitate real-time information sharing among allies to detect and neutralize cyberattacks as they happen.

In Ukraine's case, while it's not a NATO member, the alliance has provided support by sharing intelligence on cyber threats and offering technical assistance to secure critical infrastructure. This collaboration underscores the importance of international cooperation in addressing borderless threats.

Q: How does NATO coordinate with other international organizations and countries to mitigate cross-border cyber threats?

A: Coordination is essential because cyberattacks rarely stay confined within national borders. A malware attack on a single country's network can cascade into global disruptions, affecting supply chains, financial systems, and critical infrastructure worldwide. To counter this, NATO collaborates with various international organizations, including the European Union, the United Nations, and regional cybersecurity bodies.

One key area of collaboration is intelligence sharing. By pooling information about attack vectors, malware signatures, and threat actors, we can create a more comprehensive understanding of the cyber threat landscape. This shared intelligence helps nations prepare for and respond to attacks more effectively.

We also work together on joint training exercises. For example, NATO's annual Cyber Coalition exercise brings together representatives from member states, partner nations, and international organizations to simulate large-scale cyberattacks. These exercises are invaluable for testing our collective response capabilities and identifying gaps in our defenses.

Q: What role do public-private partnerships play in NATO's cyber resilience efforts?

A: Public-private partnerships are critical to cybersecurity because much of the infrastructure we aim to protect—power grids, communication networks, financial systems—is owned and operated by private companies. These organizations are often the first to detect cyber threats and play a crucial role in mitigating their impact.

At NATO, we actively engage with private sector leaders, from tech companies to cybersecurity firms, to enhance our collective defenses. For example:

- We collaborate with companies like Microsoft and Cisco to develop tools and frameworks for detecting and responding to threats.
- Through partnerships with cybersecurity firms, we gain access to advanced threat intelligence that informs our strategies.
- We also work with critical infrastructure operators to ensure they have the resources and knowledge to secure their systems against potential attacks.

These partnerships are mutually beneficial. Private companies gain insights from NATO's global threat intelligence, while we benefit from their technological innovations and expertise.

Q: Can you share any specific examples of how international cooperation has helped mitigate a significant cyber threat?

A: Certainly. A notable example is the response to the NotPetya malware attack in 2017. Although initially targeting Ukrainian systems, the malware quickly spread to multinational companies and critical infrastructure across the globe. The sheer scale of the attack required a coordinated international response.

NATO, alongside other organizations like the European Union, worked closely with member states and private sector partners to contain the spread of the malware. By sharing real-time intelligence and deploying technical expertise, we were able to assist affected nations in restoring systems and preventing further disruptions. This incident highlighted the importance of rapid information sharing and collaboration in mitigating large-scale cyber threats.

Another example is NATO's support for Ukraine during the ongoing conflict. While not directly intervening, we've provided tools, training, and intelligence to help Ukraine secure its critical infrastructure against state-sponsored cyberattacks. This has been particularly important in defending energy grids and communication networks, which have been frequent targets.

Q: What are the biggest challenges in fostering international cooperation for cyber defense?

A: One of the biggest challenges is the **diversity of capabilities** among nations. Some countries have advanced cyber defense infrastructure, while others are still developing their capacities. Bridging this gap requires significant investment in training, funding, and technical assistance.

Another challenge is **attribution**—determining who is behind a cyberattack. Without clear attribution, it's difficult to coordinate an effective international response. Adversaries often mask their activities, using proxies or false flags to create ambiguity. Developing better tools and frameworks for attribution is a key focus area for NATO and its partners.

Finally, there's the issue of **trust and information sharing**. Cybersecurity often involves sensitive intelligence, and nations may be hesitant to share information due to concerns about security breaches or political considerations. Building trust through transparency and consistent collaboration is critical for overcoming this hurdle.

Q: What lessons can the global community learn from NATO's approach to cyber defense?

A: NATO's approach highlights the importance of **proactivity** and **collaboration** in cybersecurity. A few key lessons include:

1. **Collective Defense Works**: Nations are stronger when they work

together. Pooling resources and expertise creates a more robust defense against common threats.

2. **Training and Preparedness Are Essential**: Regular simulations and exercises ensure that nations and organizations are prepared to respond to cyberattacks effectively.
3. **Public-Private Collaboration Is Vital**: Engaging the private sector helps bridge gaps in technology and intelligence, strengthening overall resilience.
4. **International Norms Are Needed**: Establishing clear rules of engagement in cyberspace can reduce the risk of escalation and promote accountability.

Q: Looking ahead, what steps must the global community take to strengthen cyber defenses?

A: The global community needs to focus on a few key areas:

1. **Invest in Capacity Building**: Nations with limited resources need support to develop their cybersecurity infrastructure and expertise. This includes funding, training, and access to technology.
2. **Establish Global Norms**: We need universally accepted norms and regulations for behavior in cyberspace, similar to the Geneva Conventions for traditional warfare.
3. **Enhance Real-Time Collaboration**: Developing platforms for real-time information sharing can help nations respond more quickly to emerging threats.
4. **Focus on Resilience**: Cyber defense isn't just about prevention—it's also about ensuring that systems can recover quickly from attacks. Investments in redundancy, backup systems, and recovery protocols are critical.

Q: What gives you hope as you work in this challenging field?

A: What gives me hope is the growing recognition of cybersecurity as a global priority. Nations are starting to understand that cyber threats are not isolated incidents—they affect everyone. The collaboration I've seen, especially during crises like the Russia-Ukraine conflict, shows that when the international community comes together, we can achieve remarkable outcomes.

I'm also encouraged by the innovation happening in this space. From AI-driven threat detection to quantum encryption, the tools we're developing are becoming more advanced every day. With continued cooperation and investment, I believe we can build a safer and more resilient digital future.

Markus Engel's insights highlight the critical role of international organizations like NATO in fostering collaboration and strengthening cyber defenses. His experiences remind us that in the face of borderless threats, collective action is our strongest asset.

9.4 Ethical Hackers: The White Hats of the Cyber Battlefield

Ethical hackers, often referred to as "white hats," play a critical role in defending against cyberattacks by identifying vulnerabilities, neutralizing threats, and strengthening security systems. In the context of the Russia-Ukraine war, ethical hackers have emerged as a key component of cyber defense, leveraging their skills to protect critical infrastructure and counteract adversarial cyber campaigns. Through organized efforts such as bug bounty programs and vulnerability disclosures, these individuals and groups have contributed significantly to cyber resilience. However, operating in wartime presents unique challenges and ethical dilemmas for ethical hackers, testing the boundaries of their roles and responsibilities.

Ethical Hackers and Their Contributions to Cyber Defense

Identifying and Mitigating Vulnerabilities: Ethical hackers use penetration testing and vulnerability assessments to identify weaknesses in systems before malicious actors can exploit them. By simulating potential attacks, they provide organizations with actionable insights to strengthen their defenses.

During the Russia-Ukraine war, ethical hackers have been instrumental in uncovering vulnerabilities in Ukrainian government networks, critical infrastructure, and communication platforms, allowing these systems to be fortified against attacks.

Counteracting Malicious Campaigns: Ethical hackers have actively disrupted cyberattacks by neutralizing malware, blocking phishing campaigns, and tracing the origins of malicious traffic. Their proactive interventions have minimized the impact of state-sponsored cyber operations targeting Ukraine.

For example, ethical hackers have worked to dismantle botnets used to launch distributed denial-of-service (DDoS) attacks, ensuring the availability of key services during critical moments.

Enhancing Threat Intelligence: By analyzing attack patterns and dissecting malware, ethical hackers contribute valuable threat intelligence that informs broader cybersecurity efforts. This intelligence helps organizations anticipate future attacks and adapt their defenses accordingly.

Ethical hackers collaborating with Ukrainian authorities and global cybersecurity firms have shared insights into the tactics, techniques, and procedures (TTPs) used by adversaries, enhancing situational awareness across the cyber battlefield.

Organized Efforts: Bug Bounty Programs and Vulnerability Disclosures

Bug Bounty Programs: Bug bounty programs incentivize ethical hackers to identify and report vulnerabilities in exchange for monetary rewards. These programs create a structured environment for ethical hacking, ensuring that discovered vulnerabilities are disclosed responsibly.

Ukrainian organizations, including government agencies and private companies, have launched bug bounty programs during the war to strengthen their cyber defenses. By crowdsourcing expertise, these programs have accelerated the discovery and remediation of vulnerabilities.

Coordinated Vulnerability Disclosures: Ethical hackers participate in coordinated vulnerability disclosure programs, working with organizations to address security gaps before making findings public. This approach minimizes the risk of exploitation while ensuring transparency.

During the conflict, ethical hackers have disclosed vulnerabilities in critical systems to Ukrainian authorities, allowing for rapid patching and mitigation.

Global Collaboration Through Platforms: Platforms like HackerOne and Bugcrowd facilitate collaboration between ethical hackers and organizations worldwide. These platforms have played a significant role in mobilizing ethical hacking talent to support Ukraine's cyber defense efforts.

By providing a centralized system for reporting and tracking vulnerabilities, these platforms ensure that ethical hackers' contributions are efficiently utilized.

Challenges and Ethical Dilemmas Faced by Ethical Hackers in Wartime

Navigating Legal and Ethical Boundaries: Operating during wartime presents unique legal and ethical challenges for ethical hackers. Actions taken to disrupt adversaries or protect systems may inadvertently cross legal boundaries or violate international norms.

For example, hacking into adversarial systems, even with the intent to neutralize threats, raises questions about accountability and the potential for escalation.

Risk of Misuse of Discovered Vulnerabilities: Ethical hackers may uncover vulnerabilities that could be exploited by malicious actors if not handled responsibly. Ensuring that discovered flaws are disclosed only to trusted parties is essential to prevent unintended consequences.

During the conflict, ethical hackers have faced the challenge of ensuring their findings are used solely for defensive purposes, particularly when collaborating with decentralized groups like hacktivists.

Pressure and Resource Constraints: Wartime conditions create intense pressure for ethical hackers to deliver rapid results, often with limited resources. Balancing speed and thoroughness is a constant challenge, as incomplete assessments can lead to overlooked vulnerabilities or unintended outcomes.

Ethical hackers supporting Ukraine have operated in high-stakes environments, where delays or missteps could have severe consequences for national security.

Personal and Professional Risks: Ethical hackers involved in wartime efforts face personal risks, including retaliation from adversaries and potential legal challenges in jurisdictions with ambiguous cyber laws.

Their association with cyber defense operations can make them targets for adversarial groups, increasing the need for operational security and anonymity.

The Broader Role of Ethical Hackers in Modern Cyber Defense

The contributions of ethical hackers during the Russia-Ukraine war illustrate their indispensable role in modern cyber defense. Their efforts have not only protected critical systems but also demonstrated the value of decentralized, community-driven approaches to cybersecurity. Ethical hackers provide:

✓ **A Proactive Line of Defense:** By identifying vulnerabilities and mitigating threats before they materialize, ethical hackers reduce the risk of catastrophic attacks.

✓ **A Bridge Between Sectors:** Ethical hackers often serve as intermediaries between private companies, governments, and global cybersecurity communities, fostering collaboration and knowledge sharing.

✓ **A Model for Resilience:** Their ability to adapt and innovate in response to evolving threats highlights the importance of flexibility and ingenuity in cyber defense.

In conclusion, ethical hackers have emerged as pivotal players on the digital battlefield, combining technical expertise with a commitment to safeguarding systems and data. Their contributions during the Russia-Ukraine war underscore the importance of integrating ethical hacking into national and organizational cybersecurity strategies. However, the challenges and ethical dilemmas they face highlight the need for clear guidelines, support frameworks, and recognition of their critical role in modern conflict. As cyber threats continue to evolve, ethical hackers will remain at the forefront of efforts to protect digital infrastructure and ensure the resilience of interconnected systems worldwide.

9.5 Building Long-Term Resilience Through Public-Private Collaboration

The evolving nature of cyber threats, as highlighted by the Russia-Ukraine war, underscores the need for sustained and strategic efforts to build long-term resilience. Public-private collaboration has emerged as a cornerstone of effective cybersecurity, leveraging the strengths of both sectors to protect critical infrastructure, enhance awareness, and prepare for future challenges. This section examines how public-private partnerships contribute to resilient infrastructure, the importance of cyber education and awareness, and the lessons from Ukraine that can inform global efforts to strengthen cybersecurity resilience.

Developing Resilient Infrastructure with Public-Private Partnerships

Pooling Resources and Expertise: Public-private partnerships (PPPs) allow governments and private entities to pool resources, expertise, and technology to fortify critical infrastructure against cyber threats. By combining the strategic oversight of governments with the technical capabilities of the private sector, PPPs create robust and adaptive defenses.

In Ukraine, partnerships with global firms such as Microsoft and Cisco provided the government with advanced tools and real-time threat intelligence, helping to secure essential systems during the conflict.

Securing Critical Infrastructure: Critical infrastructure—such as energy grids, transportation networks, and financial systems—requires coordinated efforts to mitigate vulnerabilities. Governments often lack the specialized tools and expertise to protect these systems alone, making private sector collaboration essential.

Joint initiatives, such as NATO's efforts to secure member states' cyber infrastructure through public-private partnerships, demonstrate the potential of these collaborations to address vulnerabilities at scale.

Establishing Cybersecurity Standards: PPPs play a vital role in developing and enforcing cybersecurity standards that ensure consistency across sectors. These standards include best practices for risk assessment, incident response, and system recovery, which are essential for building resilience.

For example, the National Institute of Standards and Technology (NIST) Cybersecurity Framework in the United States has been widely adopted through collaboration between public and private entities, serving as a global benchmark for cyber resilience.

Coordinating Incident Response: When cyber incidents occur, public-private collaboration enables a unified and efficient response. Governments and private companies can share intelligence, coordinate containment efforts, and recover compromised systems faster than if operating independently.

During the Russia-Ukraine war, joint response teams composed of Ukrainian authorities and private cybersecurity firms successfully mitigated the impact of ransomware attacks and other cyber threats.

Importance of Cyber Education and Awareness for Future Preparedness

Closing the Cybersecurity Skills Gap: The demand for skilled cybersecurity professionals far exceeds supply, creating vulnerabilities that adversaries can exploit. Investments in education and training are critical to developing the workforce needed to defend against evolving threats.

Ukraine's focus on training IT professionals and building capacity through initiatives like Cisco's Networking Academy helped create a talent pool capable of responding to cyber challenges during the conflict.

Raising Public Awareness: Public awareness campaigns play a crucial role in equipping individuals and organizations with the knowledge needed to recognize and respond to cyber threats. Educating citizens about phishing scams, ransomware, and basic cyber hygiene reduces the likelihood of successful attacks.

Ukraine launched national awareness programs to teach citizens how to identify disinformation and secure their personal devices, fostering a culture of vigilance and responsibility.

Integrating Cybersecurity into Education Systems: Incorporating cybersecurity into school and university curriculums ensures that future generations are prepared to navigate an increasingly digital world. Hands-on training, certifications, and competitions can inspire young talent to pursue careers in cybersecurity.

Global efforts, such as CyberPatriot competitions and partnerships with tech companies, provide models for integrating cybersecurity education into formal learning environments.

Simulated Exercises for Preparedness: Regularly conducted cyber defense exercises simulate real-world attack scenarios, testing the readiness of individuals, organizations, and governments. These exercises identify weaknesses and improve coordination among stakeholders.

NATO's Cyber Coalition exercises, which include public and private participants, offer valuable insights into defending against large-scale attacks and fostering resilience.

Lessons from Ukraine for Global Cybersecurity Resilience

The Value of Decentralized Cyber Defense: Ukraine's reliance on decentralized cyber defense efforts, including contributions from ethical hackers and volunteer groups, demonstrated the power of community-driven initiatives in responding to cyber threats.

Countries can adopt similar models by creating platforms that enable citizens and organizations to contribute to national cybersecurity efforts, such as bug bounty programs or cyber volunteer networks.

International Collaboration as a Force Multiplier: Ukraine's partnerships with allied nations, global cybersecurity firms, and international organizations underscored the importance of collaboration in building resilience. Sharing intelligence, resources, and expertise enabled Ukraine to counter sophisticated cyber threats more effectively.

The success of these partnerships highlights the need for a unified global approach to cybersecurity, where nations and corporations work together to address shared risks.

Rapid Adaptation and Innovation: The war forced Ukraine to innovate rapidly, developing custom tools and strategies to address unique challenges. This agility allowed the nation to stay ahead of adversaries and minimize disruptions.

Investing in research and development and fostering innovation ecosystems can help nations and organizations respond quickly to emerging threats.

Proactive Preparation Reduces Impact: Ukraine's investments in cybersecurity infrastructure and training before the conflict played a critical role in its ability to withstand cyberattacks. Proactive measures, such as regular security assessments and incident response planning, can significantly reduce the impact of cyber threats.

Nations and organizations worldwide should prioritize proactive preparedness to ensure resilience in the face of growing cyber risks.

Building a Resilient Future

Long-term resilience against cyber threats requires sustained collaboration between public and private sectors, continuous education and awareness efforts, and a commitment to proactive preparedness. The lessons from Ukraine illustrate how these elements can be combined to create a robust defense strategy capable of withstanding even the most sophisticated cyberattacks.

By fostering partnerships, investing in education, and leveraging global collaboration, the international community can build a more secure and resilient digital landscape, protecting critical systems and ensuring stability in an increasingly interconnected world.

Conclusion: Cyber Defense and Resilience

The Russia-Ukraine war has demonstrated the critical importance of a robust and adaptive cyber defense in modern conflicts. Ukraine's success in countering sophisticated cyberattacks highlights the power of collaboration, innovation, and preparedness in navigating the complexities of the digital battlefield. By leveraging the expertise of ethical hackers, the resources of global cybersecurity firms, and the strength of public-private partnerships, Ukraine has developed a model of resilience that offers valuable lessons for the international community.

This chapter explored the tools, strategies, and collaborative efforts that have defined Ukraine's cyber defense, from grassroots initiatives to advanced threat detection systems and international support. It also delved into the role of ethical hackers and the importance of education and awareness in building a cyber-ready workforce. These elements collectively form the foundation of a resilient digital infrastructure capable of withstanding and recovering from cyber threats.

As cyberattacks continue to evolve in scale and sophistication, the need for collective action and long-term investments in cybersecurity has never been more urgent. Ukraine's experience underscores that resilience is not just about defending against attacks but also about fostering a culture of collaboration, innovation, and adaptability. By embracing these principles, nations and organizations can better prepare for the challenges of the digital age and protect the systems that underpin global security and prosperity.

Chapter 10: The Future of Cyber Warfare

The Russia-Ukraine war has underscored the transformative role of cyber warfare in modern conflicts, signaling a new era where digital battlegrounds are as critical as physical ones. As cyberattacks continue to evolve in sophistication and scale, they pose unprecedented challenges to national security, global stability, and the integrity of digital systems. This chapter explores the future of cyber warfare, drawing lessons from the Russia-Ukraine conflict and examining the trends, ethical dilemmas, and urgent need for global regulations and norms. As nations and organizations brace for the next generation of cyber threats, understanding and addressing these challenges will be crucial in shaping a secure and cooperative digital future.

10.1 Lessons from the Russia-Ukraine Conflict

The Russia-Ukraine war has provided invaluable insights into the evolving nature of cyber warfare and the strategies required to counter its impacts. Ukraine's ability to withstand sophisticated cyberattacks offers key lessons for nations and organizations worldwide, emphasizing the importance of proactive defenses, resilience, and international collaboration. This section examines the critical takeaways from Ukraine's cyber defense strategies, highlights the role of resilience in mitigating large-scale attacks, and explores the global implications of cyber spillovers.

Key Takeaways from Ukraine's Cyber Defense Strategies

The Importance of Grassroots Cyber Efforts: Ukraine's reliance on its IT community and ethical hackers highlights the power of decentralized, grassroots efforts in cyber defense. These individuals provided real-time threat assessments, neutralized attacks, and shared intelligence, often outpacing traditional response mechanisms.

The "IT Army of Ukraine," a decentralized collective of cyber volunteers, exemplified how coordinated public participation can strengthen national cybersecurity.

Leveraging International Support: Collaboration with global cybersecurity firms and allied nations was critical in Ukraine's ability to defend against state-sponsored cyberattacks. Organizations like Microsoft, Cisco, and NATO's Cooperative Cyber Defence Centre of Excellence provided technical expertise, tools, and real-time threat intelligence.

This underscores the need for international partnerships in responding to cyber conflicts, as no nation can effectively counter these threats alone.

Integration of Cyber and Traditional Defense: Ukraine's cyber defense strategies were integrated with its overall defense efforts, showcasing the necessity of combining digital and physical responses. Cyber teams worked alongside military and intelligence units to secure critical infrastructure, protect communication systems, and mitigate the effects of hybrid warfare.

The Role of Cyber Resilience in Mitigating Large-Scale Attacks

Proactive Measures Reduce Impact: Ukraine's investments in cybersecurity before the conflict significantly reduced the impact of large-scale attacks. Regular vulnerability assessments, incident response planning, and public awareness campaigns ensured that systems and personnel were better prepared to handle threats.

Proactive defense measures, such as deploying intrusion detection systems and adopting multi-factor authentication, enabled rapid containment of cyber incidents.

The Value of Redundancy and Adaptability: Resilience was a cornerstone of Ukraine's cyber defense. By building redundant systems, such as backup communication networks, and adopting flexible strategies,

Ukraine ensured continuity of operations even during intense cyber campaigns.

For example, during DDoS attacks on government websites, alternative platforms were quickly deployed to maintain public access to critical information.

Collaboration Between Public and Private Sectors: Public-private partnerships enhanced Ukraine's ability to recover from cyberattacks. These collaborations ensured that resources, expertise, and infrastructure were shared effectively, accelerating recovery efforts and minimizing downtime.

The involvement of cybersecurity firms in mitigating attacks on energy grids and financial systems demonstrates the critical role of private entities in bolstering national resilience.

Global Implications of Cross-Border Cyber Spillovers

Interconnected Systems Amplify Risks: The NotPetya malware attack, which originated in Ukraine, serves as a stark reminder of how localized cyber conflicts can have global consequences. The malware spread rapidly across borders, disrupting multinational corporations and critical systems worldwide.

This highlights the vulnerabilities of interconnected systems and the need for international cooperation to address shared risks.

The Need for Global Cyber Norms: The global spillover effects of cyberattacks during the Russia-Ukraine war have underscored the importance of establishing international norms for responsible behavior in cyberspace. Agreements that prohibit attacks on civilian infrastructure and promote transparency are essential for reducing the risks of escalation and unintended consequences.

Raising Awareness of Global Cybersecurity Challenges: The war has brought attention to the broader challenges of cyber warfare, including the difficulty of attribution, the potential for escalation, and the ethical dilemmas posed by offensive cyber operations.

Nations and organizations worldwide must adopt lessons from Ukraine to enhance their own cybersecurity strategies, ensuring that they are prepared for the growing threat of cross-border cyber conflicts.

In summary, the lessons from the Russia-Ukraine conflict highlight the importance of resilience, collaboration, and innovation in navigating the challenges of modern cyber warfare. By investing in proactive defense measures, fostering international partnerships, and addressing the global implications of cyber spillovers, nations can build the capacity to withstand and recover from future cyber threats. Ukraine's experience offers a roadmap for strengthening cybersecurity in an increasingly interconnected and volatile digital world.

10.2 The Evolution of Cyber Threats in Modern Warfare

The digital age has redefined the nature of warfare, with cyber threats emerging as a key component of modern conflicts. The evolution of cyberattack methodologies, the integration of cyber tactics with traditional military strategies, and the rise of hybrid warfare have transformed how nations and organizations approach security. This section explores the trends shaping cyber threats in contemporary conflicts, the convergence of cyber and traditional strategies, and the implications of hybrid warfare on global stability.

Trends in Cyberattack Methodologies: AI-Driven Threats and Weaponized Malware

The Emergence of AI-Driven Cyber Threats: Artificial intelligence (AI) has become a double-edged sword in cybersecurity, enabling both defenders and attackers to enhance their capabilities. AI-driven cyber threats use machine learning algorithms to adapt, evade detection, and target vulnerabilities more effectively.

Examples include:

✓ **AI-Powered Phishing:** Attackers use AI to craft highly personalized phishing emails that are more convincing and difficult to identify.

✓ **Automated Malware:** AI enables malware to evolve autonomously, bypassing traditional defense mechanisms.

✓ **Adversarial AI:** Cybercriminals use adversarial AI to manipulate machine learning models in critical systems, such as facial recognition or fraud detection algorithms.

Weaponized Malware and Its Increasing Sophistication: Malware remains a cornerstone of cyber warfare, with state-sponsored groups developing sophisticated tools to disrupt, disable, and destroy targets.

Key Examples:

✓ *Wiper Malware:* Used extensively during the Russia-Ukraine war, this type of malware erases data on infected systems, rendering them inoperable.

✓ *Ransomware as a Weapon:* Attackers increasingly deploy ransomware not only for financial gain but also to destabilize adversaries by paralyzing critical infrastructure.

✓ *Supply Chain Attacks:* Malware is embedded in trusted software or hardware, allowing attackers to compromise multiple systems through a single entry point, as seen in the SolarWinds breach.

Zero-Day Exploits and Advanced Persistent Threats (APTs): Attackers are increasingly leveraging zero-day vulnerabilities—flaws in software that are unknown to the vendor—for stealthy, high-impact operations.

APT groups, often linked to state actors, conduct long-term campaigns to infiltrate networks, gather intelligence, and sabotage critical systems.

The Convergence of Cyber and Traditional Military Strategies

Cyber Operations as a Force Multiplier: Cyber tactics are now fully integrated into traditional military strategies, serving as a force multiplier to enhance the effectiveness of physical operations.

Examples include:

✓ **Disabling Enemy Infrastructure:** Cyberattacks on power grids and communication networks weaken an adversary's ability to coordinate military responses.

✓ **Psychological Operations:** Disinformation campaigns amplified through social media create confusion and erode morale.

✓ **Intelligence Gathering:** Cyber espionage provides critical insights into enemy strategies, troop movements, and supply chains.

Pre-Conflict Cyber Attacks: Cyber operations often precede physical conflicts, softening targets and creating strategic advantages. For instance, cyberattacks targeting Ukrainian infrastructure began well before the physical invasion, disrupting energy grids, financial systems, and government networks.

These preemptive strikes blur the lines between war and peace, complicating responses and raising the stakes of escalation.

Integration with Autonomous Systems: The convergence of cyber capabilities with autonomous systems, such as drones and unmanned vehicles, has added a new dimension to warfare. Cyberattacks on these systems can disable or hijack them, turning advanced technology into a liability.

The Rise of Hybrid Warfare and Its Impact on Global Security

What Is Hybrid Warfare? Hybrid warfare combines conventional military tactics with cyber operations, disinformation campaigns, economic coercion, and other non-traditional methods to achieve strategic objectives.

The goal is to destabilize adversaries by exploiting vulnerabilities across multiple domains, creating confusion and delaying effective responses.

The Role of Cyber in Hybrid Conflicts: Cyber operations are central to hybrid warfare, amplifying its impact by targeting critical systems, spreading propaganda, and undermining trust in institutions.

✓ During the Russia-Ukraine war, hybrid tactics included:

✓ Cyberattacks on government networks and power grids.

✓ Disinformation campaigns aimed at eroding international support for Ukraine.

✓ Economic sabotage through ransomware attacks on financial systems.

Challenges in Responding to Hybrid Threats: Hybrid warfare creates ambiguity, making it difficult to attribute attacks or determine when a conflict has begun. This ambiguity complicates decision-making for policymakers and military leaders.

The decentralized nature of cyber operations, often involving non-state actors and hacktivist groups, further blurs the lines of accountability.

Global Implications of Hybrid Warfare: Hybrid conflicts have far-reaching implications for global security, as their effects often extend beyond the immediate theater of operations. For example:

✓ Disinformation campaigns can polarize societies and influence elections in distant nations.

✓ Cyberattacks on multinational corporations disrupt global supply chains and economies.

✓ The risk of escalation increases as hybrid tactics provoke retaliation across multiple domains.

In summary, the evolution of cyber threats in modern warfare has transformed the nature of conflict, creating new challenges for national security and global stability. The integration of cyber tactics with traditional military strategies and the rise of hybrid warfare have demonstrated the need for adaptive, multi-layered defenses. As AI-driven threats and weaponized malware continue to advance, nations and organizations must invest in proactive measures to anticipate and counter emerging risks. The lessons from the Russia-Ukraine conflict serve as a blueprint for understanding and addressing these evolving challenges in an increasingly interconnected and volatile world.

10.3 The Need for Global Regulations and Cyber Norms

The proliferation of cyber warfare and the increasing complexity of digital threats have highlighted the urgent need for global regulations and norms to govern behavior in cyberspace. The lack of clear rules for cyber conflict has created a gray area where actors operate with relative impunity, complicating efforts to ensure accountability and prevent escalation. This section explores the challenges of attribution and accountability, examines proposals for international cyber norms, and underscores the role of multilateral organizations in fostering cooperation and compliance.

Addressing the Challenges of Attribution and Accountability in Cyberspace

The Difficulty of Attribution: Cyberattacks often originate from anonymous or obfuscated sources, making it challenging to identify the perpetrators with certainty. Hackers use techniques such as proxy servers, encryption, and hijacked infrastructure to mask their activities, complicating forensic investigations.

State-sponsored actors frequently operate under the guise of independent groups or cybercriminals, further muddying the waters. For example, Russian-linked groups like Fancy Bear have conducted sophisticated operations that are difficult to trace definitively to state sponsorship.

The Risk of Misattribution: Incorrectly attributing a cyberattack can lead to severe consequences, including diplomatic fallout, unjustified retaliation, and the escalation of conflicts. This risk makes nations hesitant to act decisively, even when they suspect a specific actor.

The challenge of attribution is compounded during hybrid warfare, where cyber operations are part of a broader strategy that includes disinformation and economic coercion.

Lack of Enforcement Mechanisms: Even when perpetrators are identified, there are limited mechanisms to hold them accountable. Unlike traditional warfare, cyberspace lacks a universally recognized framework for imposing consequences on violators.

Efforts to address this gap require international cooperation to establish legal and operational frameworks for investigating and prosecuting cyberattacks.

Proposals for Establishing International Norms for Cyber Behavior

Defining Acceptable Conduct in Cyberspace: International norms for cyber behavior should establish clear guidelines for what constitutes acceptable and unacceptable actions in cyberspace. These norms could include:

✓ Prohibiting attacks on civilian infrastructure, such as hospitals, energy grids, and communication networks.

✓ Banning the use of ransomware or wiper malware to target critical systems.

✓ Refraining from interfering in democratic processes, including elections and referendums.

Building on Existing Frameworks: Efforts to establish cyber norms can draw from existing frameworks, such as the United Nations Group of Governmental Experts (UNGGE) recommendations, which emphasize state responsibility and the protection of critical infrastructure.

The Tallinn Manual, developed by legal scholars and military experts, provides a comprehensive guide to the application of international law in cyberspace and offers a foundation for building consensus on cyber norms.

Incentivizing Compliance: To encourage adherence to cyber norms, nations must see tangible benefits, such as improved security, economic stability, and access to shared intelligence. Mechanisms for enforcing compliance could include:

✓ Diplomatic pressure and sanctions against violators.

✓ Rewarding nations that adopt and uphold cyber norms with technical assistance and capacity-building programs.

✓ Establishing independent bodies to investigate and attribute cyberattacks, similar to the International Atomic Energy Agency (IAEA) for nuclear compliance.

The Role of Multilateral Organizations in Fostering Cooperation and Compliance

The United Nations as a Convening Platform: The United Nations (UN) provides a unique platform for fostering dialogue and building consensus on cyber norms. Initiatives such as the Open-Ended Working Group (OEWG) and the UNGGE facilitate discussions among member states to address emerging threats and promote responsible behavior in cyberspace.

The UN's inclusive approach ensures that both developed and developing nations have a voice in shaping the rules of cyberspace, addressing concerns about equity and access.

Regional Organizations' Contributions: Regional organizations, such as the European Union (EU), ASEAN, and the African Union (AU), play a complementary role in advancing cybersecurity norms. These groups address region-specific challenges and foster collaboration among neighboring states.

The EU's General Data Protection Regulation (GDPR) serves as a model for harmonizing data protection standards, while initiatives like NATO's Cooperative Cyber Defence Centre of Excellence focus on enhancing collective defense.

Engaging the Private Sector and Civil Society: Multilateral organizations must engage private companies, cybersecurity firms, and civil society to ensure that norms are comprehensive and enforceable. The private sector, which owns and operates much of the world's critical infrastructure, has a vested interest in supporting robust cybersecurity frameworks.

Civil society organizations, including advocacy groups and academic institutions, provide valuable perspectives on the ethical implications of cyber governance and advocate for the protection of human rights in digital spaces.

Establishing Accountability Mechanisms: Multilateral organizations can play a pivotal role in establishing mechanisms for accountability, such as:

✓ An international cyber tribunal to investigate and adjudicate cyber incidents.

✓ Independent attribution bodies to provide credible assessments of cyberattacks.

✓ Collaborative agreements for joint responses to large-scale cyber incidents, ensuring a coordinated and effective approach.

Path Forward: Building a Cooperative Digital Future

Promoting Transparency and Trust: Transparency in cyber operations, such as disclosing vulnerabilities and sharing incident reports, can reduce suspicion and foster trust among nations. Collaborative exercises, such as cyber defense simulations, can enhance preparedness and build confidence in shared capabilities.

Harmonizing Global Standards: Aligning national and regional cybersecurity standards with global norms ensures consistency and reduces vulnerabilities in interconnected systems. International standards bodies, such as ISO, play a critical role in this effort.

Educating Stakeholders on Cyber Norms: Raising awareness among governments, businesses, and individuals about the importance of cyber norms and regulations can drive compliance and build a culture of accountability.

In conclusion, the need for global regulations and cyber norms has never been more urgent. The Russia-Ukraine war has demonstrated the devastating potential of unregulated cyber conflict and the importance of establishing clear rules for behavior in cyberspace. By addressing the challenges of attribution and accountability, fostering cooperation through multilateral organizations, and incentivizing compliance, the global community can build a safer and more stable digital environment. In an era where cyberspace is both a battleground and a shared resource, the pursuit of collective security and ethical governance must remain a priority.

10.4 Preparing for the Next Generation of Cyber Warfare

The increasing sophistication of cyber threats demands a forward-looking approach to security. Preparing for the next generation of cyber warfare involves investing in proactive defense strategies, fostering global capacity through education and collaboration, and strengthening international alliances to address shared risks. As the digital battlefield evolves, nations and organizations must adopt comprehensive strategies to anticipate and counter emerging threats effectively. This section examines these critical elements and their role in shaping a resilient and secure future.

Investing in Proactive Defense Strategies and Advanced Technologies

Adopting a Proactive Security Mindset: Reactive responses to cyber threats are no longer sufficient. Organizations and governments must adopt a proactive security approach, emphasizing prevention, early detection, and rapid containment of cyber incidents.

Proactive strategies include continuous monitoring of network traffic, threat hunting, and vulnerability management to identify potential risks before they are exploited.

Harnessing Artificial Intelligence (AI) and Machine Learning (ML): AI and ML technologies are revolutionizing cyber defense by enabling faster detection of anomalies, predicting attack patterns, and automating responses. These tools help organizations stay ahead of increasingly sophisticated attackers.

Examples of AI-driven applications include:

✓ Behavioral analytics to identify insider threats.

✓ Automated incident response systems to isolate infected devices.
✓ Predictive modeling to anticipate emerging attack vectors.

Strengthening Critical Infrastructure: Protecting critical infrastructure, such as energy grids, transportation systems, and financial networks, requires targeted investments in advanced security solutions.

Technologies such as quantum encryption, zero-trust architectures, and blockchain-based security protocols are being developed to enhance the resilience of essential systems.

Enhancing Supply Chain Security: Cyberattacks on supply chains have highlighted the need for greater transparency and security throughout the lifecycle of products and services. Organizations must implement stringent vetting processes, regular audits, and secure communication channels to protect against supply chain vulnerabilities.

Developing Cyber-Resilient Systems: Resilience is critical for ensuring continuity during cyber incidents. This includes implementing robust backup systems, disaster recovery plans, and redundant infrastructure to minimize disruption and ensure rapid recovery.

Building Global Capacity Through Education, Training, and Collaboration

Addressing the Cybersecurity Skills Gap: The demand for skilled cybersecurity professionals far exceeds supply, creating vulnerabilities that adversaries can exploit. Investing in education and training is essential to develop the workforce needed to secure the digital landscape.

Initiatives such as scholarships, certification programs, and public-private partnerships can encourage more individuals to pursue careers in cybersecurity.

Integrating Cybersecurity into Formal Education: Introducing cybersecurity concepts in school curriculums and higher education fosters a culture of awareness and preparedness from an early age. Programs should include hands-on training, ethical hacking courses, and opportunities to participate in cybersecurity competitions.

For example, initiatives like CyberPatriot and Cybersecurity Bootcamps provide students with real-world experience and prepare them for careers in the field.

Promoting Collaboration Across Sectors: Cross-sector collaboration between governments, private companies, academia, and non-profit organizations is essential for building a comprehensive cybersecurity ecosystem.

Forums like the World Economic Forum's Centre for Cybersecurity facilitate dialogue and knowledge-sharing, enabling stakeholders to address challenges collectively.

Expanding Global Knowledge Sharing: Cyber threats are a shared challenge, requiring nations to share intelligence, best practices, and technical expertise. Platforms like NATO's Cooperative Cyber Defence Centre of Excellence (CCDCOE) and the Global Forum on Cyber Expertise (GFCE) play a pivotal role in promoting international collaboration.

Conducting Regular Cyber Defense Exercises: Simulated exercises, such as war games and incident response drills, help organizations and governments test their preparedness for cyberattacks. These exercises identify weaknesses, improve coordination, and ensure readiness for real-world scenarios.

Initiatives like NATO's Cyber Coalition exercise demonstrate the value of collaborative simulations in building capacity and fostering resilience.

Strengthening International Alliances to Counter Shared Threats

The Importance of Collective Security: Cyber threats often transcend borders, making international alliances essential for addressing shared risks. Collaborative frameworks allow nations to coordinate defenses, share intelligence, and respond collectively to global cyber incidents.

Examples of such alliances include NATO's Article 5, which recognizes cyberattacks as potential triggers for collective defense, and regional partnerships like the European Union's Cybersecurity Strategy.

Establishing Joint Response Mechanisms: Rapid and coordinated responses to cyber incidents are critical for minimizing damage and preventing escalation. International alliances can establish joint response mechanisms, such as shared threat intelligence platforms and cyber incident response teams.

During the Russia-Ukraine war, allied nations provided Ukraine with technical support, real-time intelligence, and resources to counter cyberattacks, demonstrating the value of collective action.

Aligning National and International Policies: Harmonizing cybersecurity policies and standards across nations reduces inconsistencies and strengthens global defenses. This includes aligning national laws on data protection, cybercrime, and critical infrastructure security with international frameworks.

Efforts like the Budapest Convention on Cybercrime and the Tallinn Manual on the International Law Applicable to Cyber Operations provide blueprints for achieving policy alignment.

Fostering Trust Among Allies: Building trust among nations is a cornerstone of effective alliances. Transparency in cyber operations, such as disclosing vulnerabilities and incident reports, fosters confidence and reduces the risk of misunderstandings or misattribution.

Joint training programs, exchange programs for cybersecurity professionals, and collaborative research initiatives further strengthen trust and cooperation.

Deterring Adversaries Through United Fronts: Strong alliances serve as a deterrent to adversaries by demonstrating collective strength and resolve. Coordinated responses to cyberattacks, such as imposing sanctions or launching countermeasures, reinforce the message that malicious actions will have consequences.

In summary, preparing for the next generation of cyber warfare requires a holistic approach that integrates proactive defense strategies, workforce development, and international collaboration. By investing in advanced technologies, building global capacity through education and training, and strengthening alliances, nations and organizations can anticipate and counter emerging threats. The digital battlefield is constantly evolving, but with coordinated efforts and forward-looking strategies, the global community can create a resilient and secure future. The lessons from the

Russia-Ukraine conflict serve as a reminder of the importance of preparedness and unity in navigating the complexities of cyber warfare.

10.5 The Ethical Challenges of Cyber Warfare

Cyber warfare, while providing new tools for national defense and conflict resolution, raises significant ethical dilemmas. The use of offensive cyber capabilities, the manipulation of information through disinformation campaigns, and the evolving role of ethical hackers challenge existing norms and principles in warfare. As cyber operations increasingly influence global security and societal dynamics, understanding and addressing these ethical challenges is critical. This section examines the balance between offensive capabilities and ethical considerations, the societal impact of disinformation and psychological operations, and the future role of ethical hacking in mitigating cyber conflicts.

Balancing Offensive Capabilities with Ethical Considerations

The Justification for Offensive Cyber Operations: Offensive cyber capabilities allow nations to preemptively neutralize threats, disrupt enemy operations, and protect critical infrastructure. These capabilities are increasingly viewed as essential components of national defense strategies.

Examples include disabling adversary communication networks, deploying ransomware to cripple logistics, or sabotaging weapons systems to prevent their use. However, such actions carry significant ethical and legal implications.

Collateral Damage in Cyberspace: Offensive cyber operations often have unintended consequences, impacting civilians and neutral entities. For example:

✓ A cyberattack on an energy grid can cause widespread blackouts, affecting hospitals, businesses, and homes.

✓ Malware designed for a specific target can spread globally, as seen in the NotPetya attack, which disrupted systems far beyond its intended scope.

These risks highlight the need for strict ethical guidelines to minimize collateral damage.

Defining Proportionality and Necessity: Ethical frameworks for cyber warfare must address the principles of proportionality (ensuring the response is commensurate with the threat) and necessity (limiting actions to those essential for achieving objectives).

Determining what constitutes a proportional or necessary response in cyberspace is complex, given the lack of physical damage and the difficulty in assessing long-term impacts.

Accountability and Transparency: Unlike traditional military actions, cyber operations often lack transparency, making it difficult to hold actors accountable. Clear protocols and international agreements are needed to govern the use of offensive cyber capabilities and ensure accountability for misuse.

The Impact of Disinformation and Psychological Operations on Society

The Weaponization of Information: Disinformation campaigns and psychological operations (PsyOps) are key components of modern cyber warfare, aiming to manipulate public opinion, destabilize societies, and erode trust in institutions.

Examples include spreading false narratives about military actions, influencing elections through fake news, and creating panic through fabricated crises.

Erosion of Trust and Democratic Integrity: The widespread use of disinformation undermines trust in governments, media, and democratic processes. This erosion of trust weakens social cohesion and makes societies more vulnerable to further manipulation.

For instance, during the Russia-Ukraine war, disinformation campaigns targeted both domestic and international audiences, spreading confusion and polarizing communities.

The Ethical Dilemma of Psychological Operations: PsyOps blur the line between legitimate strategic communication and unethical manipulation. While nations argue that these tactics are necessary for national security, their long-term effects on societal stability and individual autonomy raise ethical concerns.

The use of deepfakes, AI-generated content, and social media bots further complicates the ethical landscape, making it harder to distinguish truth from fabrication.

Mitigating the Impact of Disinformation: Countering disinformation requires a multi-faceted approach, including media literacy programs, fact-checking initiatives, and regulatory measures to hold platforms accountable for spreading false information.

Ethical considerations must guide these efforts to ensure that countermeasures do not infringe on freedom of expression or other fundamental rights.

The Future Role of Ethical Hacking in Mitigating Cyber Conflicts

Ethical Hackers as Defenders of Digital Integrity: Ethical hackers, or "white hats," play a crucial role in identifying vulnerabilities, securing systems, and mitigating the impact of cyberattacks. Their proactive efforts help prevent conflicts and protect critical infrastructure.

During the Russia-Ukraine war, ethical hackers were instrumental in detecting malware, neutralizing threats, and sharing intelligence with affected organizations.

Organized Efforts to Enhance Cybersecurity: Ethical hacking is increasingly institutionalized through bug bounty programs, penetration testing initiatives, and vulnerability disclosure platforms. These programs incentivize ethical hackers to collaborate with organizations and governments in strengthening defenses.

Global platforms like HackerOne and Bugcrowd enable ethical hackers to contribute to cybersecurity efforts while adhering to strict ethical guidelines.

Challenges Faced by Ethical Hackers: Ethical hackers often navigate a fine line between legitimate actions and potential legal or ethical violations. For example:

✓ Accessing systems without explicit authorization, even for defensive purposes, may breach legal frameworks.

✓ The disclosure of vulnerabilities must be handled responsibly to prevent exploitation by malicious actors.

Clear guidelines and legal protections are needed to empower ethical hackers while ensuring their actions align with ethical principles.

Promoting Ethical Standards in Cyber Defense: The integration of ethical hacking into national cybersecurity strategies requires the establishment of clear standards and training programs. Ethical hackers should receive support and recognition for their contributions to global cyber resilience.

Collaboration between ethical hackers, governments, and private organizations can create a unified approach to addressing cyber conflicts.

Balancing Ethics and Efficacy in Cyber Warfare

Developing Ethical Frameworks for Cyber Operations: Nations must develop and adhere to ethical frameworks that guide the conduct of cyber warfare, ensuring that actions align with international law and humanitarian principles.

These frameworks should address:

✓ The proportionality and necessity of offensive operations.
✓ Accountability for collateral damage.
✓ The ethical use of disinformation and psychological tactics.

Promoting International Collaboration: Addressing the ethical challenges of cyber warfare requires global cooperation. Multilateral agreements can establish norms for responsible behavior, ensuring that cyber operations do not undermine societal stability or violate human rights.

Platforms like the United Nations Group of Governmental Experts (UNGGE) and the Tallinn Manual provide valuable foundations for developing these norms.

Educating Stakeholders on Ethical Cyber Practices: Governments, organizations, and individuals must be educated on the ethical implications of cyber warfare. This includes training military personnel on the ethical use of cyber capabilities and raising public awareness about the impact of disinformation and PsyOps.

In summary, the ethical challenges of cyber warfare demand careful consideration as nations and organizations navigate the complexities of the digital battlefield. Balancing offensive capabilities with ethical principles, mitigating the societal impact of disinformation, and empowering ethical hackers to defend against threats are critical steps in addressing these dilemmas. By fostering international collaboration, developing clear ethical frameworks, and prioritizing transparency and accountability, the global community can ensure that cyber operations uphold the principles of justice, integrity, and respect for human rights in an increasingly interconnected world.

Interview with Dr. Aisha Raman, AI and Cyber Ethics Researcher

Dr. Aisha Raman is a leading researcher in artificial intelligence and cyber ethics, specializing in the intersection of emerging technologies and their societal impact. Her work explores the ethical considerations of AI-driven cyber capabilities, focusing on accountability, transparency, and the challenges of maintaining ethical standards in cyber warfare.

Q: Dr. Raman, thank you for joining us. Let's start with the big question: What are the key ethical dilemmas surrounding the use of cyber capabilities in warfare?

A: Thank you for having me. The ethical dilemmas in cyber warfare are vast and complex, largely because the digital domain blurs many of the lines that traditional warfare operates within. One of the primary dilemmas is the issue of **proportionality**. In physical warfare, proportionality means ensuring that the response to an attack is measured and does not cause excessive harm. In cyberspace, this is far harder to define. For instance, a cyberattack on a power grid could disrupt hospitals and water systems, leading to humanitarian crises far beyond the intended target.

Another key dilemma is **attribution**—determining who is responsible for a cyberattack. In many cases, cyberattacks are designed to obscure the attacker's identity, often through proxies or false flags. Without clear attribution, responses to cyberattacks can risk targeting the wrong entities, escalating conflicts unnecessarily.

Finally, there's the question of **accountability**. When an AI-driven system launches a cyberattack or neutralizes a threat, who is responsible for its actions? The programmer? The military commander? Or the government that deployed the system? This lack of clarity creates significant ethical and legal challenges.

Q: How do AI and automation complicate the ethical landscape of cyber warfare?

A: AI and automation are game-changers in cyber warfare, but they come with significant ethical complexities. On one hand, AI-driven systems can process vast amounts of data, identify threats, and respond far more quickly than humans ever could. This speed and efficiency are critical in defending against sophisticated, rapidly evolving cyberattacks.

However, the same capabilities introduce risks. AI systems operate based on algorithms and training data, which may include biases or incomplete information. This can lead to unintended consequences. For example:

- An AI system might misclassify legitimate traffic as malicious, disrupting essential services.
- In offensive operations, AI might escalate a situation by launching disproportionate counterattacks or targeting unintended systems.

Another concern is **autonomy**. Fully autonomous cyber systems raise the possibility of decisions being made without human oversight. While this might improve reaction times, it also removes a layer of human judgment, increasing the risk of errors or ethical breaches.

Q: Offensive cyber capabilities often involve preemptive actions. How do we balance the need for proactive defense with ethical responsibility?

A: This is one of the most pressing ethical questions in cyber warfare. Preemptive actions—such as disabling a potential adversary's cyber infrastructure before an attack occurs—are inherently controversial. On the one hand, they can prevent harm by neutralizing threats early. On the other, they risk violating sovereignty and escalating conflicts.

Balancing these considerations requires a framework that prioritizes **transparency, accountability, and proportionality**. Before taking preemptive action, decision-makers must weigh several factors:

1. **Certainty of the Threat**: Is there clear evidence that an attack is imminent, or is the action based on assumptions?
2. **Potential Collateral Damage**: Could the action inadvertently harm civilians or critical infrastructure unrelated to the conflict?
3. **Legal and Ethical Standards**: Does the action comply with international laws and norms, or does it set a dangerous precedent?

Preemptive actions should always involve human oversight and should be limited to situations where the evidence is unequivocal, and the benefits clearly outweigh the risks.

Q: What role does international law play in addressing the ethical challenges of cyber warfare?

A: International law is critical, but it's also struggling to keep pace with the rapid evolution of technology. Existing frameworks, such as the Geneva Conventions, were designed for physical warfare and don't fully address the unique challenges of cyberspace.

There have been efforts to adapt international law for cyber warfare. For example, the **Tallinn Manual** provides guidance on how international law applies to cyber operations. It addresses issues like sovereignty, proportionality, and state responsibility in cyberspace. However, it's not a legally binding document, and its adoption varies widely.

To move forward, the international community needs to establish **clear and enforceable norms for cyber behavior**, including:

- Prohibiting attacks on civilian infrastructure like hospitals, water systems, and energy grids.
- Requiring transparency in the development and deployment of AI-driven cyber tools.
- Creating accountability mechanisms to hold states and organizations responsible for violations.

Collaboration is key. Nations, international organizations, and the private sector must work together to build consensus and establish these norms.

Q: What practical steps can be taken to improve accountability in cyber warfare?

A: Improving accountability requires action on multiple levels:

1. **Developing Clear Command Structures**: In cyber operations, it must be explicitly clear who is responsible for each decision. This includes delineating accountability for AI-driven systems, where human oversight must remain central.
2. **Transparency in Cyber Capabilities**: While secrecy is often necessary in military operations, transparency about the rules governing cyber warfare can build trust and reduce misunderstandings. For instance, states could commit to publicly outlining the ethical standards guiding their cyber operations.
3. **Independent Oversight Bodies**: Establishing international oversight bodies to investigate cyber incidents and ensure compliance with established norms would be a significant step forward. These bodies could function similarly to the International Atomic Energy Agency (IAEA) in the nuclear domain.
4. **Mandatory Auditing and Reporting**: Governments and organizations deploying offensive cyber tools should be required

to audit their operations regularly and report on their compliance with ethical and legal standards.

Q: Do you see a path forward for balancing innovation with ethical responsibility in cyber warfare?

A: I do, but it requires a collective effort. Innovation in cyber capabilities is inevitable, and we shouldn't shy away from leveraging technology to enhance security. However, this innovation must be guided by ethical principles. Governments, developers, and international organizations need to collaborate to ensure that the pursuit of technological advancement doesn't come at the expense of accountability, humanity, and global stability.

The path forward involves investing in **ethical design practices**, where cybersecurity tools are developed with built-in safeguards to prevent misuse. It also means fostering a culture of responsibility, where decision-makers at all levels—from developers to military leaders—are held accountable for their actions.

Finally, education plays a key role. By training the next generation of cyber professionals in both technical and ethical disciplines, we can ensure that future innovations align with our shared values.

Q: What gives you hope as you work on these challenging issues?

A: What gives me hope is the growing awareness of the need for ethical frameworks in cyber warfare. More and more, I see governments, organizations, and individuals recognizing the importance of accountability and transparency in this space. The fact that these conversations are happening now, rather than after catastrophic failures, is a positive sign.

I'm also encouraged by the potential of international cooperation. While progress can be slow, the groundwork is being laid for meaningful agreements on cyber norms. If we continue to prioritize collaboration over competition, I believe we can navigate these challenges and create a digital landscape that is both secure and ethical.

Dr. Aisha Raman's insights shed light on the intricate ethical challenges of cyber warfare. Her work emphasizes the importance of balancing innovation with responsibility, reminding us that accountability and transparency are not just ideals—they are essential for ensuring a secure and just digital future.

10.6 A Vision for a Secure Digital Future

As cyber threats continue to evolve in scale and complexity, envisioning a secure digital future has become a global imperative. The interconnectedness of today's systems, economies, and societies means that no single nation or organization can address cybersecurity challenges alone. Collective action, ethical alignment of technological innovation, and strong global leadership are essential to safeguarding digital ecosystems and ensuring a resilient, inclusive future. This section explores the importance of collaborative resilience-building, the integration of ethics into technological advancements, and the role of global leadership in fostering a secure digital landscape for all.

The Importance of Collective Action to Build Resilience

Cybersecurity as a Shared Responsibility: The digital realm transcends borders, making cybersecurity a shared responsibility among governments, private organizations, and individuals. Collaborative efforts to secure interconnected systems are critical for addressing threats that impact global stability and security.

The Russia-Ukraine war demonstrated how cross-border cyberattacks ripple through industries and economies, highlighting the need for coordinated defenses to protect critical infrastructure and digital assets.

Public-Private Partnerships for Enhanced Security: Public-private partnerships (PPPs) are vital for leveraging the strengths of both sectors to build cyber resilience. Governments provide strategic oversight, while private companies contribute technological expertise and resources.

Examples include the involvement of global cybersecurity firms like Microsoft and Cisco in protecting Ukraine's networks, which showcased how collective efforts mitigate risks during conflicts.

Knowledge Sharing and Capacity Building: Sharing threat intelligence, best practices, and technical expertise among nations and organizations strengthens collective defenses. Platforms like the Cyber Threat Alliance and NATO's Cooperative Cyber Defence Centre of

Excellence (CCDCOE) facilitate this collaboration, ensuring preparedness and rapid response to emerging threats.

Capacity-building initiatives, particularly in developing nations, ensure that all stakeholders have the tools and expertise needed to participate in global cybersecurity efforts.

International Cooperation for Incident Response: Coordinated responses to major cyber incidents are essential for minimizing damage and preventing escalation. Establishing international response teams, shared communication platforms, and joint simulations can enhance global readiness.

The Global Forum on Cyber Expertise (GFCE) exemplifies how multilateral cooperation addresses cyber challenges collectively.

Aligning Technological Innovation with Ethical Responsibility

The Dual Role of Technology in Cybersecurity: Technological innovation is both a solution and a challenge in cybersecurity. While advanced tools like AI and blockchain enhance defenses, they also introduce new vulnerabilities and ethical dilemmas. Balancing innovation with responsibility is critical to ensuring long-term security.

For instance, AI-driven threat detection systems offer unparalleled accuracy, but adversarial AI tactics, such as deepfake creation, pose significant risks.

Embedding Ethics in Design and Deployment: Ethical considerations must guide the development and deployment of emerging technologies to prevent misuse and ensure fairness. This includes:

✓ Designing AI algorithms that avoid biases and protect privacy.

✓ Implementing robust security measures in IoT devices to prevent exploitation.

✓ Ensuring transparency and accountability in the use of surveillance tools.

Ethical frameworks, such as those proposed by the IEEE and the European Union's AI Act, provide guidelines for aligning technological advancements with societal values.

Addressing the Ethics of Cyber Operations: Offensive cyber capabilities raise ethical questions about proportionality, collateral damage, and accountability. Establishing clear norms for the use of these capabilities ensures that actions align with international law and humanitarian principles.

Nations must commit to ethical cyber practices that prioritize stability and minimize harm to civilians and neutral entities.

Promoting Responsible Innovation: Encouraging responsible innovation involves fostering a culture of security and ethics among researchers, developers, and organizations. Public awareness campaigns, industry standards, and academic programs focused on cybersecurity ethics can drive this cultural shift.

The Role of Global Leadership in Ensuring Cybersecurity for All

Setting the Agenda for Cybersecurity: Global leadership is essential for prioritizing cybersecurity on the international agenda. Organizations like the United Nations, the World Economic Forum, and regional bodies such as the European Union must lead efforts to establish norms, coordinate responses, and mobilize resources.

Multilateral agreements, such as the Budapest Convention on Cybercrime, provide a foundation for collective action, but broader participation and enforcement are needed.

Fostering Inclusive Participation: Effective global leadership ensures that all nations, regardless of size or resources, have a voice in shaping the future of cybersecurity. Inclusivity is critical for addressing disparities in cybersecurity capabilities and ensuring that developing nations can contribute to and benefit from global efforts.

Initiatives like the Digital Development Partnership (DDP) by the World Bank support capacity building in underserved regions, fostering a more equitable digital future.

Building Trust Through Transparency: Trust is a cornerstone of international cooperation. Transparent cyber practices, such as disclosing vulnerabilities and incident reports, build confidence among nations and stakeholders.

Joint training programs, cyber defense exercises, and exchange programs for cybersecurity professionals can further strengthen trust and collaboration.

Deterring Adversaries Through Unified Action: Strong global leadership sends a clear message to malicious actors: cyberattacks will not go unchallenged. Coordinated responses, including sanctions, indictments, and diplomatic measures, deter adversaries and reinforce the collective resolve of the international community.

In summary, a secure digital future depends on collective action, ethical responsibility, and effective global leadership. By building resilience through collaboration, aligning technological advancements with societal values, and fostering inclusive participation, the international community can navigate the challenges of the evolving cyber landscape.

The lessons from the Russia-Ukraine war underscore the urgency of these efforts, highlighting both the risks of unregulated cyberspace and the potential of coordinated action to create a safer, more stable digital world. Achieving this vision requires a shared commitment to innovation, accountability, and cooperation, ensuring that cybersecurity serves as a foundation for global prosperity and peace.

Conclusion: The Future of Cyber Warfare

The Russia-Ukraine war has marked a pivotal moment in the evolution of cyber warfare, revealing both the vulnerabilities of interconnected systems and the strategic importance of cyber capabilities in modern conflicts. As cyber threats continue to grow in sophistication and reach, the lessons learned from this conflict highlight the need for global collaboration, ethical accountability, and proactive defense strategies.

The future of cyber warfare will be shaped by advances in technology, the integration of cyber and traditional military strategies, and the rise of hybrid tactics. Addressing these challenges requires the establishment of clear norms and regulations to govern behavior in cyberspace, balancing innovation with responsibility, and fostering international alliances to counter shared threats.

A secure digital future depends on collective action and visionary leadership. Nations, organizations, and individuals must work together to build resilience, promote ethical practices, and ensure that technological advancements serve the greater good. By investing in education, fostering collaboration, and aligning innovation with societal values, the global community can prepare for the complexities of cyber warfare and create a stable, secure digital landscape for generations to come. The path forward demands vigilance, unity, and a commitment to safeguarding the systems that underpin our interconnected world.

Part III: Interconnected Realities

Chapter 11: Environmental and Cyber Intersections

The intersection of cyber warfare and environmental risks represents an emerging frontier in modern conflicts. As critical infrastructure becomes increasingly digitized, cyberattacks targeting power grids, water systems, dams, and industrial facilities have the potential to cause significant ecological damage. The cascading effects of these attacks can disrupt ecosystems, harm public health, and exacerbate climate challenges. This chapter examines how cyber warfare can trigger environmental disasters, explores the vulnerabilities of eco-sensitive systems, and analyzes real-world case studies that illustrate the profound cross-domain impacts of cyber conflicts. Understanding these intersections is essential for developing strategies that protect both digital and environmental security in an interconnected world.

11.1 Cyberattacks as Catalysts for Environmental Damage

As cyberattacks target critical infrastructure, their consequences often extend beyond digital disruption to significant environmental harm. Power grids, dams, industrial facilities, and other vital systems are increasingly at risk from sophisticated cyber threats. These systems play essential roles in maintaining ecological balance and public safety, and their compromise can result in cascading failures that amplify damage across multiple domains. This section explores how cyberattacks exploit vulnerabilities in infrastructure, examines the potential for cascading failures, and provides examples of incidents with environmental repercussions.

Exploiting Vulnerabilities in Power Grids, Dams, and Industrial Facilities

Power Grids as Prime Targets: Power grids are among the most targeted infrastructures in cyber warfare due to their critical role in supporting energy, healthcare, communication, and industrial systems. A successful attack on a grid can lead to widespread blackouts, disrupting essential services and causing ecological harm.

For example, blackouts force a reliance on backup generators, which often run on fossil fuels, increasing carbon emissions and air pollution. Extended outages can also interrupt water treatment facilities, leading to untreated water entering natural ecosystems.

Dams and Water Management Systems: Dams and reservoirs are vital for water supply, flood control, and energy generation. Cyberattacks on these systems can result in catastrophic failures, such as uncontrolled water releases or dam collapses, threatening both human and ecological systems.

Hackers exploiting vulnerabilities in dam control systems could trigger flooding, eroding habitats, destroying farmland, and contaminating water supplies with sediment and pollutants.

Industrial Facilities and Chemical Plants: Industrial and chemical facilities are attractive targets for cyberattacks due to their reliance on digital control systems. A cyber-induced failure in such facilities could result in toxic leaks, fires, or explosions, with devastating environmental consequences.

For instance, tampering with sensors or valves at a chemical plant could release hazardous substances into the air or water, impacting ecosystems and public health for years.

The Potential for Cascading Failures in Critical Infrastructure

Interconnected Systems Amplify Risks: Modern infrastructure systems are deeply interconnected, meaning that an attack on one component can have ripple effects across others. For example:

✓ A cyberattack on a power grid can disrupt energy supplies to water treatment plants, leading to untreated sewage discharge into rivers.

✓ Disruptions in transportation networks caused by power outages can delay the response to environmental emergencies, worsening the impact.

Feedback Loops of Damage: Cascading failures can create feedback loops, where initial disruptions trigger secondary crises. For instance, a blackout caused by a cyberattack may shut down refrigeration systems at food storage facilities, leading to spoilage and waste that require costly and environmentally harmful disposal methods.

Economic and Ecological Consequences: Cascading failures often lead to significant economic losses and long-term ecological harm. For example, prolonged power outages can hinder industrial productivity, increase reliance on carbon-intensive energy sources, and damage biodiversity in affected regions.

Examples of Cyber Incidents with Environmental Repercussions

The Ukraine Power Grid Attacks (2015–2016): In 2015 and 2016, cyberattacks on Ukraine's power grid caused widespread blackouts, affecting hundreds of thousands of people. While the immediate impact was limited to energy disruption, these attacks exposed the vulnerabilities of critical systems and the potential for broader environmental harm.

Blackouts forced the use of diesel generators, increasing air pollution and greenhouse gas emissions. Additionally, water and sewage systems were temporarily compromised, risking ecological contamination.

Stuxnet and Industrial Sabotage (2010): The Stuxnet worm, a cyberweapon targeting Iranian nuclear facilities, highlighted the risks of cyberattacks on industrial systems. While its environmental impact was minimal, the incident demonstrated how cyberattacks could disrupt industrial processes, with potential for severe ecological consequences in facilities such as chemical plants or refineries.

The Colonial Pipeline Ransomware Attack (2021): A ransomware attack on the Colonial Pipeline in the United States disrupted fuel supplies across the East Coast, leading to panic buying and increased transportation of fuel by road. This not only increased greenhouse gas emissions but also heightened the risk of spills and accidents during fuel transport.

The incident illustrated how cyberattacks on energy infrastructure can indirectly exacerbate environmental risks by altering supply chains and increasing reliance on less sustainable practices.

Hypothetical Scenarios with Dams and Water Systems: While no large-scale cyberattacks on dams have yet occurred, hypothetical scenarios provide a glimpse of the potential devastation:

- ✓ A cyberattack causing the release of water from a dam could lead to flooding, destroying ecosystems, displacing wildlife, and eroding soil.
- ✓ Tampering with water treatment facilities could contaminate drinking water supplies, impacting both human health and aquatic ecosystems.

In conclusion, cyberattacks targeting critical infrastructure pose significant risks to the environment, with cascading failures amplifying their impact. From power grids and dams to industrial facilities, the vulnerabilities of these systems highlight the urgent need for robust cybersecurity measures that integrate environmental considerations. As the intersection of cyber warfare and environmental risks becomes increasingly evident, protecting eco-sensitive infrastructure must be a global priority. Recognizing these threats and learning from past incidents will be essential in mitigating future risks and safeguarding the delicate balance between technological advancement and ecological stability.

11.2 Risks of Digitally-Induced Ecological Disasters

Cyberattacks targeting critical infrastructure have the potential to trigger catastrophic ecological disasters, affecting ecosystems, human health, and the economy. From energy systems to water management and industrial plants, the reliance on digital controls makes these systems vulnerable to cyber threats with far-reaching environmental implications. This section explores the risks associated with cyberattacks on energy systems, water management facilities, and manufacturing and chemical plants, highlighting the potential for significant environmental damage.

Targeting Energy Systems: From Blackouts to Oil Spills

The Ripple Effects of Power Grid Attacks: Cyberattacks on power grids can result in widespread blackouts, disrupting essential services and increasing reliance on fossil fuels. These disruptions often lead to environmental consequences, including:

✓ Increased emissions from diesel generators, which are deployed to provide emergency power.

✓ Interruptions in renewable energy operations, as grid instability can force wind and solar plants offline.

Prolonged outages may also hinder the operation of environmental monitoring systems, delaying responses to ecological crises.

Oil and Gas Infrastructure at Risk: Energy pipelines, refineries, and storage facilities are attractive targets for cyberattacks. Compromising these systems can lead to:

✓ **Oil Spills:** A cyberattack manipulating pipeline controls could cause ruptures, resulting in oil spills that contaminate soil, water, and marine ecosystems.

✓ **Fires and Explosions:** Disabling safety systems at refineries or storage facilities increases the risk of fires and explosions, releasing toxic substances into the air and water.

For example, the Colonial Pipeline ransomware attack in 2021 disrupted fuel supplies, forcing increased reliance on road transport, which heightened the risk of spills and increased greenhouse gas emissions.

Nuclear Energy Facilities: Cyberattacks on nuclear power plants pose a unique and severe ecological threat. While direct attacks on reactor operations are rare, targeting auxiliary systems such as cooling or waste management could lead to catastrophic outcomes, including radioactive contamination.

Although no major incidents have occurred, scenarios like these highlight the need for robust cybersecurity measures in the energy sector.

Cyber Threats to Water Management Systems and Their Ecological Impact

Tampering with Dams and Reservoirs: Cyberattacks on dams and reservoirs can result in unauthorized water releases or dam failures, leading to:

✓ **Flooding:** Uncontrolled water releases can destroy ecosystems, erode soil, and displace wildlife and human communities.

✓ **Habitat Destruction:** Floods caused by dam breaches can submerge forests, wetlands, and agricultural land, leading to long-term ecological damage.

Such attacks could also exacerbate drought conditions by draining reservoirs critical for water supply.

Contamination of Water Supplies: Water treatment facilities rely on automated control systems to monitor and manage chemical levels. Cyberattacks manipulating these systems can cause:

✓ **Overdosing or Underdosing of Chemicals:** Excessive chemicals, such as chlorine, can harm aquatic ecosystems, while insufficient treatment may lead to the spread of waterborne diseases.

✓ **Introduction of Contaminants:** Hackers could introduce harmful substances into drinking water systems, affecting public health and ecological balance.

In 2021, a cyberattack on a water treatment facility in Oldsmar, Florida, attempted to increase the sodium hydroxide concentration to dangerous levels, highlighting the potential risks.

Impact on Irrigation and Agriculture: Cyberattacks on irrigation systems can disrupt water distribution, leading to overwatering or drought conditions. This not only affects crop yields but also depletes water resources and harms soil health.

The ripple effects of such disruptions can impact food security and the environment simultaneously.

The Environmental Cost of Attacks on Manufacturing and Chemical Plants

Chemical Plant Vulnerabilities: Manufacturing facilities and chemical plants are prime targets for cyberattacks due to their reliance on industrial control systems (ICS). Compromising these systems can result in:

✓ **Toxic Leaks:** Cyberattacks can manipulate storage and processing equipment, causing the release of hazardous chemicals into the environment.

✓ **Fires and Explosions:** Disabling safety systems increases the risk of catastrophic accidents, releasing harmful pollutants into the air and water.

For example, a breach in a fertilizer plant's systems could lead to ammonia leaks, devastating local ecosystems and endangering nearby communities.

Disruptions in Production Chains: Attacks on manufacturing plants can halt production, leading to increased waste and resource inefficiencies. Perishable materials may spoil, requiring environmentally costly disposal methods such as incineration or landfill use.

Waste Mismanagement: Cyberattacks disrupting waste management processes can lead to improper disposal of industrial byproducts, contaminating land and water sources.

If automated systems for handling hazardous waste are compromised, the resulting pollution can have long-lasting ecological consequences.

Global Supply Chain Impacts: Disruptions in manufacturing due to cyberattacks can ripple through global supply chains, forcing industries to source materials from less sustainable alternatives or transport goods over longer distances, increasing carbon footprints.

In summary, Cyberattacks targeting energy systems, water management facilities, and industrial plants highlight the profound environmental risks posed by modern digital threats. These incidents underscore the need for integrating cybersecurity measures with environmental protection strategies to prevent ecological disasters. As the reliance on automated systems grows, securing these infrastructures against cyber threats must become a global priority to safeguard ecosystems and public health. Recognizing the interconnectedness of cyber and environmental risks is essential for building a sustainable and secure future.

11.3 Case Studies: Cross-Domain Impacts of Cyber Warfare on the Environment

Cyber warfare's reach extends beyond digital systems, creating ripple effects that impact the environment, economies, and societies. Specific incidents have highlighted the interplay between cyberattacks and ecological harm, serving as cautionary tales of the vulnerabilities in critical infrastructure. This section explores real-world case studies such as the Ukraine power grid attacks and the Colonial Pipeline incident, as well as hypothetical scenarios involving dams and water systems, to illustrate the cross-domain impacts of cyber warfare on the environment.

The Ukraine Power Grid Attacks: Linking Blackouts to Environmental Stress

Background of the Attacks: In 2015 and 2016, Ukraine's power grid was targeted by state-sponsored cyberattacks, leading to widespread blackouts in the Ivano-Frankivsk region and Kyiv. These incidents were among the first examples of cyber operations directly disrupting a nation's energy infrastructure.

The attackers used sophisticated malware, including BlackEnergy and Industroyer, to gain access to control systems, causing power outages that affected hundreds of thousands of people.

Environmental Consequences: The blackouts forced the use of backup power systems, primarily diesel generators, which increased air pollution and carbon emissions. Diesel generators emit particulate matter and greenhouse gases, contributing to poor air quality and climate change.

Disrupted power supplies also impacted water treatment facilities, leading to untreated wastewater being discharged into rivers and streams. This contamination posed risks to aquatic ecosystems and public health.

Cascading Effects on Ecosystems: The temporary loss of power affected industries dependent on stable electricity, such as agriculture and food processing. Interruptions in refrigeration and irrigation systems resulted in spoilage and water wastage, exacerbating resource inefficiencies.

Wildlife habitats reliant on environmental monitoring systems were left vulnerable due to the failure of sensors and communication networks.

Lessons Learned: The Ukraine power grid attacks demonstrated how cyber warfare can amplify environmental stress by disrupting interconnected systems. Strengthening cybersecurity in energy infrastructure is essential to mitigate such risks and protect ecosystems from collateral damage.

The Colonial Pipeline Incident: The Ripple Effects on Energy and Ecosystems

Overview of the Incident: In 2021, a ransomware attack on the Colonial Pipeline, the largest refined petroleum pipeline in the United States, forced the shutdown of its operations for several days. The disruption caused widespread fuel shortages across the East Coast and prompted panic buying.

The attack, attributed to the DarkSide ransomware group, highlighted vulnerabilities in critical energy infrastructure and the cascading effects of cyberattacks.

Environmental Impacts: The pipeline's shutdown increased reliance on road and rail transport for fuel delivery, leading to higher greenhouse gas emissions and increased risk of spills during transit.

Panic buying resulted in hoarding and improper storage of fuel, raising the likelihood of leaks and accidental fires, both of which have long-term environmental repercussions.

Disruptions in fuel supply also affected emergency response operations, delaying actions needed to address environmental hazards such as chemical spills or forest fires.

Broader Ecological Concerns: The incident underscored the fragility of energy supply chains and their environmental implications. Any prolonged disruption to pipelines can significantly impact air and water quality, as well as wildlife dependent on unaffected habitats.

The event also highlighted the indirect ecological costs of cyberattacks, as disrupted logistics often lead to less sustainable alternatives being deployed.

Key Takeaways: The Colonial Pipeline incident revealed the interconnected risks of cyber warfare and environmental harm, emphasizing the need for robust cybersecurity measures in the energy sector to protect ecosystems and ensure sustainability.

Potential Threats to Dams and Water Systems: Examining Hypothetical Scenarios

Cyberattacks on Dams and Reservoirs: Dams and reservoirs are vital for water supply, flood control, and energy generation. A cyberattack compromising their control systems could lead to catastrophic outcomes, such as:

✓ **Uncontrolled Water Releases:** Triggering flash floods that destroy habitats, erode soil, and displace communities.

✓ **Dam Failures:** Resulting in massive flooding, submerging ecosystems, and contaminating water sources with sediment and pollutants.

Impact on Ecosystems and Biodiversity: Flooding caused by dam breaches can wipe out vegetation, displace wildlife, and disrupt aquatic ecosystems. The sudden change in water flow can also harm fish populations, especially species dependent on specific flow rates for spawning.

Contaminated floodwaters can spread pollutants, affecting both terrestrial and aquatic life for years.

Contamination of Drinking Water Supplies: Water treatment plants are increasingly reliant on digital systems to manage chemical levels and filtration processes. A cyberattack manipulating these controls could result in:

✓ **Over-Treatment:** Excessive chlorine or other chemicals harming aquatic ecosystems.

✓ **Under-Treatment:** Allowing harmful bacteria or toxins to enter drinking water supplies, endangering public health and aquatic habitats.

Case Example: Oldsmar Water Treatment Facility (2021): In a notable incident, a hacker gained access to the Oldsmar water treatment facility in Florida and attempted to increase sodium hydroxide levels to dangerous concentrations. Although the attack was thwarted, it demonstrated the potential for cyberattacks to harm both humans and ecosystems through water contamination.

Need for Cybersecurity in Water Management: Securing water infrastructure against cyber threats is critical to prevent ecological disasters. This includes regular assessments of vulnerabilities, real-time monitoring systems, and incident response protocols tailored to environmental risks.

In conclusion, these case studies highlight the tangible and hypothetical risks posed by cyberattacks to the environment. Whether targeting energy grids, pipelines, or water systems, the cross-domain impacts of cyber warfare emphasize the interconnectedness of digital and ecological systems. Learning from these incidents and preparing for potential scenarios is essential to mitigating the environmental consequences of future cyber conflicts. Strengthening cybersecurity in eco-sensitive infrastructure and integrating environmental considerations into cyber defense strategies will be key to ensuring resilience in an increasingly digital and interconnected world.

11.4 Bridging Cybersecurity and Environmental Protection

As the intersection of cyber threats and environmental risks becomes more evident, integrating environmental considerations into cybersecurity strategies is essential. Safeguarding eco-sensitive infrastructure requires collaboration between governments, private organizations, and international institutions to address vulnerabilities and prevent ecologically harmful cyberattacks. This section explores the need to incorporate environmental risks into cybersecurity planning, highlights collaborative approaches for protecting critical systems, and examines the role of international norms in fostering responsible cyber behavior.

Integrating Environmental Risks into Cybersecurity Strategies

Recognizing the Environmental Impacts of Cyberattacks: Cyberattacks targeting critical infrastructure, such as power grids, dams, and industrial plants, often result in cascading failures with environmental consequences.

Incorporating environmental risk assessments into cybersecurity planning ensures that potential ecological impacts are considered during incident response and mitigation efforts.

Developing Eco-Sensitive Cyber Protocols: Cybersecurity strategies should include protocols specifically designed to protect eco-sensitive infrastructure. This includes:

✓ Regular audits to identify vulnerabilities in systems that control water management, energy distribution, and chemical processing.

✓ Real-time monitoring of critical systems to detect anomalies that could lead to ecological damage.

✓ Contingency plans to minimize environmental harm in the event of a successful attack.

Cross-Disciplinary Collaboration: Cybersecurity teams should work closely with environmental scientists and engineers to understand the potential ecological implications of system disruptions.

For example, collaboration can help prioritize the protection of infrastructure with the highest environmental stakes, such as water treatment plants or nuclear facilities.

Investing in Resilient Infrastructure: Building redundancy and resilience into critical infrastructure reduces the likelihood of cascading failures. For instance:

✓ Backup power systems for water treatment facilities can ensure continued operation during a cyber-induced blackout.

✓ Automating safety shutdown mechanisms in industrial plants can prevent chemical leaks or explosions during cyber incidents.

Collaborative Approaches to Safeguard Eco-Sensitive Infrastructure

Public-Private Partnerships (PPPs): Public-private collaboration is critical for protecting eco-sensitive infrastructure, much of which is owned and operated by private entities.

Governments and private organizations can work together to:

✓ Share intelligence on emerging cyber threats.

✓ Develop standardized protocols for securing infrastructure.

✓ Conduct joint simulations and training exercises focused on eco-sensitive scenarios.

Community Involvement and Awareness: Engaging local communities in cybersecurity efforts increases vigilance and awareness around eco-sensitive infrastructure. For example:

✓ Community-led initiatives to monitor water quality or energy usage can complement digital monitoring systems.

✓ Public education campaigns can teach residents how to recognize and report suspicious cyber activity that could impact local ecosystems.

International Collaboration: Protecting transboundary infrastructure, such as shared rivers, dams, and pipelines, requires coordinated efforts between nations.

Regional agreements, such as the Danube River Protection Convention, can incorporate cybersecurity into broader environmental protection frameworks.

Leveraging Technology and Innovation: Advanced technologies, such as AI and IoT, can enhance the security of eco-sensitive systems. For instance:

✓ AI-powered analytics can detect and neutralize cyber threats in real time.

✓ IoT sensors can provide continuous data on water quality, energy flows, and emissions, enabling rapid response to anomalies.

The Role of International Norms in Preventing Environmentally Harmful Cyberattacks

Establishing Global Standards for Cybersecurity: International norms and agreements can define acceptable behavior in cyberspace, prohibiting attacks that target eco-sensitive infrastructure.

Existing frameworks, such as the United Nations Group of Governmental Experts (UNGGE) recommendations, provide a foundation for developing these norms.

Prohibiting Attacks on Critical Environmental Systems: Specific protections for infrastructure critical to environmental sustainability, such as water treatment plants and renewable energy facilities, should be codified in international law.

These protections can be modeled after the Geneva Conventions, which prohibit attacks on civilian infrastructure during armed conflicts.

Encouraging Transparency and Accountability: Nations and organizations should commit to transparent practices, such as disclosing vulnerabilities and sharing incident reports, to build trust and facilitate joint efforts.

Accountability mechanisms, such as independent attribution bodies, can ensure that perpetrators of environmentally harmful cyberattacks are identified and held responsible.

Incentivizing Compliance Through Collaboration: International organizations, such as the United Nations and the World Economic Forum, can incentivize nations to adopt and comply with eco-sensitive cyber norms by providing technical assistance, funding, and capacity-building programs.

Promoting Cross-Border Cybersecurity Agreements: Regional agreements, such as the European Union's Cybersecurity Strategy, demonstrate the potential for collaborative frameworks to address shared threats.

Expanding these agreements to include environmental considerations can enhance their relevance and effectiveness.

In conclusion, bridging cybersecurity and environmental protection is essential for safeguarding eco-sensitive infrastructure in an increasingly interconnected world. By integrating environmental risks into cybersecurity strategies, fostering collaborative approaches, and establishing international norms, nations and organizations can mitigate the ecological impacts of cyber warfare. The intersection of digital and environmental security requires innovative solutions and coordinated efforts, ensuring that critical systems are protected not only against cyber threats but also for the sustainability of ecosystems and communities worldwide.

Interview with Natalia Zhukova, NGO Worker Specializing in Post-Conflict Environmental Restoration

Natalia Zhukova is a project coordinator with Green Horizons, an international NGO dedicated to environmental restoration in conflict zones. Based in Ukraine, Natalia has been on the frontlines of post-conflict recovery, working to rehabilitate damaged ecosystems and secure eco-sensitive infrastructure in war-affected areas.

Q: Natalia, thank you for speaking with us. Can you begin by describing the challenges you've faced in restoring ecosystems in conflict zones like Ukraine?

A: Thank you for having me. Restoring ecosystems in conflict zones is a complex and often overwhelming task. The damage is multifaceted—bombings destroy forests, industrial spills contaminate soil and water, and entire ecosystems are disrupted. On top of that, the ongoing conflict creates safety challenges. Many areas remain inaccessible due to unexploded ordnance, landmines, or active military operations.

Another challenge is the **lack of baseline data**. Before we can start restoration, we need to assess the extent of the damage, but war often disrupts environmental monitoring systems, leaving us with incomplete information. For example, in the Donbas region, much of the environmental monitoring infrastructure was destroyed, making it difficult to measure the full impact of pollution from damaged industrial facilities.

Finally, there's the issue of resources. Environmental restoration is expensive, and funding is often limited in post-conflict settings where humanitarian needs take priority. Balancing these immediate needs with long-term ecological recovery is one of our biggest challenges.

Q: Can you share specific examples of environmental restoration projects you've worked on in Ukraine?

A: Certainly. One project we've been involved in is the cleanup and restoration of the Dnipro River Basin. During the conflict, several industrial facilities near the river were damaged, leading to chemical spills that contaminated the water and surrounding soil. Our team worked with local authorities and international partners to contain the pollution using barriers and bio-absorbent materials. We also introduced phytoremediation techniques—using plants to absorb toxins from the soil and water.

Another significant project is reforesting parts of the Donetsk region. Forests here have been devastated by bombings and fires caused by military activity. We partnered with local communities to plant native tree species, focusing on re-establishing biodiversity and stabilizing the soil to prevent erosion.

We're also working on a pilot project to restore wetlands near the Black Sea, which have been affected by oil spills and military debris. These wetlands are crucial for local wildlife and act as natural flood barriers, so their recovery is a high priority.

Q: How do you integrate cybersecurity into your environmental restoration efforts, especially for critical infrastructure?

A: Cybersecurity has become increasingly important in our work because much of the infrastructure we're trying to protect—like water treatment plants and energy grids—is digitized and vulnerable to cyberattacks. A cyber breach in one of these systems can undo months of restoration efforts.

For example, we're working with local governments to secure the control systems of water treatment plants. This involves implementing basic cyber hygiene practices, such as regular software updates and multi-factor authentication, as well as more advanced measures like intrusion detection systems.

We also collaborate with international cybersecurity experts to train local staff. Many of these facilities were not designed with cybersecurity in mind, so building awareness and capacity is crucial. The integration of cybersecurity with environmental restoration is still a relatively new concept, but it's becoming essential as conflicts increasingly target digital infrastructure.

Q: What role does local community involvement play in your restoration projects?

A: Community involvement is absolutely central to what we do. Local communities are the stewards of the ecosystems we're trying to restore, and their participation ensures that our efforts are sustainable.

In the reforestation project I mentioned earlier, we worked closely with local residents to select tree species that would thrive in the area and meet community needs. Involving locals in planting and maintenance not only fosters a sense of ownership but also creates economic opportunities in areas where livelihoods have been disrupted by war.

Education is another key component. We run workshops to teach communities about the importance of biodiversity, sustainable farming practices, and how to monitor and protect their local environment. These efforts help ensure that the restored ecosystems are cared for long after our projects are complete.

Q: How do international organizations and governments support your work?

A: International organizations and governments play a critical role in providing the resources and expertise we need to scale our efforts. For instance, the United Nations Environment Programme (UNEP) has been a valuable partner, offering technical support and funding for large-scale projects like the Dnipro River cleanup.

We've also received support from European Union initiatives focused on sustainable development and climate resilience. These programs provide grants and facilitate knowledge sharing between NGOs, governments, and private sector experts.

Collaboration with private companies has also been impactful. For example, we've worked with tech firms to deploy drones for environmental monitoring and with chemical companies to develop advanced remediation techniques. However, more long-term funding and coordination are needed to address the scale of the challenges we're facing.

Q: What gives you hope as you work in these challenging conditions?

A: What gives me hope is the resilience of both nature and people. Despite the devastation, I've seen ecosystems begin to recover when given the chance. Wetlands that were heavily polluted are showing signs of life again, and forests are slowly regenerating.

I'm also inspired by the dedication of the people I work with, from local community members to international experts. Their commitment to protecting and restoring the environment, even in the face of immense challenges, is a constant source of motivation.

Lastly, the growing global awareness of the environmental costs of war gives me hope. More governments, organizations, and individuals are recognizing the importance of integrating environmental restoration into post-conflict recovery. This shift in mindset is crucial for creating a sustainable and peaceful future.

Q: What message would you like to share with the global community about post-conflict environmental restoration?

A: My message is that environmental restoration is not just an afterthought—it's a vital part of rebuilding communities and ensuring long-term stability. War doesn't just destroy cities and economies; it also devastates the natural systems that sustain life. If we neglect these systems, the consequences will ripple far beyond the conflict zone, affecting global ecosystems and future generations.

I urge the global community to prioritize funding and support for environmental restoration in post-conflict regions. It's not just about healing the land—it's about healing the people who depend on it.

Natalia Zhukova's insights reveal the critical role of environmental restoration in post-conflict recovery. Her work underscores the importance of integrating community involvement, international collaboration, and cybersecurity into efforts to rebuild ecosystems and secure critical infrastructure.

11.5 Preparing for Cross-Domain Threats

The increasing intersection of cyber warfare and environmental risks demands proactive preparation to address cross-domain threats effectively. Building resilience in critical infrastructure, raising awareness about the environmental implications of cyber warfare, and fostering global cooperation are essential steps to mitigate the impact of these interconnected challenges. This section explores how to develop eco-secure infrastructure, the importance of public and institutional awareness, and the need for international collaboration to reduce the risks associated with cross-domain threats.

Developing Resilience in Infrastructure with Eco-Security in Mind

Designing Eco-Secure Infrastructure: Critical infrastructure, such as energy grids, water systems, and chemical plants, must be designed with both cybersecurity and environmental risks in mind. This includes:

✓ Implementing **redundancy** in key systems to ensure continued operation during cyberattacks.

✓ Deploying **fail-safe mechanisms**, such as automated shutdowns, to prevent ecological harm during disruptions.

✓ Using **secure-by-design principles**, where cybersecurity is integrated into infrastructure development from the outset.

Integrating Environmental and Cyber Risk Assessments: Infrastructure risk assessments should evaluate both cyber vulnerabilities and environmental consequences, identifying potential cascading effects.

For example, a cyberattack on a water treatment plant could lead to ecological contamination. Proactive assessments can help prioritize safeguards for eco-sensitive components of the system.

Harnessing Technology for Real-Time Monitoring and Response: Advanced technologies, such as AI and IoT, can enhance resilience by enabling real-time monitoring and rapid response to cyber incidents:

✓ **AI-powered analytics** can predict potential cyber threats and simulate their ecological impacts, helping prioritize mitigation efforts.

✓ **IoT sensors** in energy grids or water systems can detect anomalies early, reducing the risk of cascading failures.

Digital twins—virtual replicas of physical systems—can be used to test the impact of potential cyberattacks and design more resilient systems.

Cross-Sector Collaboration for Resilience Building: Governments, private companies, and environmental organizations must work together to build infrastructure that is resilient to both cyber threats and ecological risks. For instance:

Collaborative planning between energy providers and environmental agencies ensures that both cyber and environmental factors are addressed.

Industry standards, like those developed by the International Electrotechnical Commission (IEC), can guide the integration of cyber and eco-security measures.

Raising Awareness About the Environmental Implications of Cyber Warfare

Educating Stakeholders About Cross-Domain Risks: Awareness campaigns targeting policymakers, industry leaders, and the public are essential to highlight the environmental implications of cyberattacks. Key initiatives include:

✓ Workshops and training programs for industry professionals on mitigating cross-domain risks.

✓ Public education campaigns about the potential ecological harm of cyber incidents, such as water contamination or oil spills.

Incorporating Cross-Domain Risks into Cybersecurity Training: Cybersecurity training programs should include modules on environmental risks, emphasizing the interconnectedness of digital and ecological systems.

For example, cybersecurity professionals should understand how disruptions in industrial control systems (ICS) could lead to chemical leaks or habitat destruction.

Engaging Media and Civil Society: Media and civil society organizations play a crucial role in raising awareness of cross-domain threats. Documentaries, investigative journalism, and advocacy campaigns can bring attention to the risks of cyber warfare on the environment.

Highlighting real-world incidents, such as the Colonial Pipeline attack, helps illustrate the tangible consequences of these threats.

Promoting Cross-Disciplinary Research: Encouraging collaboration between cybersecurity experts, environmental scientists, and engineers fosters innovative solutions to cross-domain challenges.

Academic institutions and think tanks can provide research and recommendations for integrating environmental considerations into cybersecurity policies.

Case for Global Cooperation to Mitigate Cross-Domain Risks

The Need for International Collaboration: Cyberattacks and environmental risks are transboundary issues that require coordinated international responses.

Shared systems, such as rivers, pipelines, and power grids, necessitate joint efforts to secure critical infrastructure and prevent cascading failures across borders.

Developing Global Frameworks for Eco-Secure Cyber Norms: International norms should explicitly address environmentally harmful cyberattacks, prohibiting actions that could lead to ecological disasters.

- ✓ Frameworks like the United Nations Group of Governmental Experts (UNGGE) recommendations can be expanded to include eco-security considerations.

- ✓ Binding agreements, similar to the Geneva Conventions, can formalize protections for eco-sensitive infrastructure during conflicts.

Regional and Bilateral Agreements: Nations sharing critical resources, such as rivers or energy grids, should establish regional or bilateral agreements to coordinate cybersecurity and environmental protection efforts.

For example, cross-border frameworks for managing shared water systems can include cybersecurity protocols to prevent tampering with dams or treatment facilities.

Resource Sharing and Capacity Building: Advanced nations must assist developing countries in building resilience against cross-domain threats. This includes:

✓ Providing technical expertise and cybersecurity tools to secure eco-sensitive infrastructure.

✓ Funding initiatives to upgrade outdated systems vulnerable to both cyber and environmental risks.

✓ Establishing regional centers of excellence for training and research on cross-domain resilience.

Conducting Joint Simulations and Exercises: International organizations, such as NATO or the Global Forum on Cyber Expertise (GFCE), can organize joint exercises simulating cross-domain attacks. These exercises:

✓ Test the readiness of nations and organizations to respond to complex threats.

✓ Identify weaknesses in coordination and communication.

✓ Foster trust and collaboration among participating stakeholders.

In summary, preparing for cross-domain threats requires an integrated approach that addresses the interconnected vulnerabilities of digital and environmental systems. By developing eco-secure infrastructure, raising awareness of the environmental implications of cyber warfare, and fostering global cooperation, nations and organizations can mitigate the risks associated with these complex challenges.

As cyber and ecological risks become increasingly intertwined, proactive measures will be essential to safeguarding both digital infrastructure and the natural environment. Global collaboration, innovation, and education will play critical roles in ensuring a secure and sustainable future for ecosystems, economies, and societies alike.

Conclusion: Environmental and Cyber Intersections

The intersection of cyber threats and environmental risks has emerged as a critical frontier in modern security. As cyberattacks increasingly target eco-sensitive infrastructure, their potential to cause ecological disasters underscores the urgency of integrating environmental considerations into cybersecurity strategies. From the Ukraine power grid attacks to the Colonial Pipeline incident, these cases highlight how vulnerabilities in digital systems can have cascading impacts on ecosystems, public health, and economies.

Preparing for cross-domain threats requires a comprehensive approach that includes developing eco-resilient infrastructure, raising awareness about the environmental implications of cyber warfare, and fostering international collaboration to address shared challenges. Establishing global norms that explicitly prohibit environmentally harmful cyberattacks is a crucial step toward protecting both the digital and natural worlds.

By recognizing the interconnected nature of cyber and environmental risks, nations and organizations can work together to safeguard critical systems and mitigate the consequences of future cyber incidents. The lessons from this chapter emphasize the importance of proactive planning, innovative solutions, and collective action in building a secure and sustainable future for all.

Chapter 12: The Role of Technology in Mitigation

As the world grapples with the aftermath of the Russia-Ukraine war, the potential of technology to address its environmental and infrastructural consequences has never been more evident. The war's impact on ecosystems, critical infrastructure, and resource security has created an urgent need for innovative solutions that go beyond traditional recovery methods. Technology offers a pathway to not only repair the damage but also to rebuild with resilience and sustainability in mind. From advanced monitoring tools and AI-driven analytics to renewable energy systems and ecological restoration techniques, technological advancements are reshaping how nations recover from conflict. This chapter explores the transformative role of technology in mitigating environmental damage, restoring ecosystems, and fostering a future rooted in sustainability and resilience.

Introduction: The Promise of Technology in Post-War Recovery

The devastation caused by the Russia-Ukraine war has left indelible marks on the environment, infrastructure, and societal systems. From polluted rivers and scorched forests to the destabilization of essential services, the scale of destruction demands innovative and sustainable solutions. While earlier chapters have detailed the environmental impacts of the conflict and the vulnerabilities it exposed, the focus now shifts to how technology can be harnessed to address these challenges.

Technology offers a transformative promise in post-war recovery, bridging the gap between understanding damage and implementing solutions. Tools such as AI-powered analytics, IoT-enabled monitoring systems, and advanced ecological restoration techniques provide not just efficiency but also precision in tackling complex problems. These innovations enable nations to rebuild with resilience, ensuring that recovery efforts are not only effective but also sustainable in the face of future risks.

As this chapter unfolds, it delves into the actionable solutions technology offers to heal war-torn environments and create a foundation for long-term recovery and stability. By leveraging advancements in science and engineering, post-war recovery can transcend mere restoration, setting a precedent for building a more sustainable and resilient future.

12.1 Advancements in Environmental Monitoring

In the wake of conflict, understanding the extent and nature of environmental damage is a critical first step toward recovery. Advancements in environmental monitoring technologies have revolutionized this process, providing tools that are both precise and efficient. Remote sensing technologies, such as satellite imagery and drones, enable comprehensive mapping and monitoring of damaged ecosystems, offering a bird's-eye view of areas impacted by war. These innovations not only reveal the immediate consequences of conflict but also provide the data needed to guide effective restoration efforts.

Satellite imagery has become an indispensable tool for assessing large-scale environmental damage. Geographic Information Systems (GIS) enhance the utility of this data, enabling the identification of deforested regions, polluted waterways, and disrupted habitats. For instance, satellites can detect changes in vegetation cover, track oil spills, or assess sediment buildup in rivers, offering valuable insights into how ecosystems are shifting in response to human activity and natural forces.

Drones complement satellite imagery by providing high-resolution visuals of specific areas, particularly those that are difficult to access on the ground. Equipped with thermal imaging capabilities, drones can identify habitat loss, monitor wildlife populations, and even detect illegal activities like logging or poaching in real-time. This technology is invaluable for creating detailed assessments of environmental conditions, informing targeted interventions, and tracking recovery progress.

The integration of AI-powered data analysis further elevates the effectiveness of environmental monitoring. Machine learning algorithms can process vast amounts of data from remote sensing tools, identifying patterns and predicting recovery timelines with unprecedented accuracy. For example, AI models can forecast how long it will take for a deforested area to regenerate naturally or determine the best locations for reforestation projects based on soil health and climate conditions.

These technologies not only streamline the monitoring process but also optimize resource allocation. By pinpointing the most critical areas for intervention, they ensure that recovery efforts are both efficient and impactful. In the context of the Russia-Ukraine war, these advancements are not just tools of observation—they are catalysts for action, enabling a data-driven approach to healing the scars left on the environment. As monitoring technologies continue to evolve, their role in post-war recovery will only grow, setting a new standard for precision and sustainability in addressing ecological crises.

12.2 Soil Remediation and Restoration Technologies

The conflict in Ukraine has left vast swaths of land contaminated by heavy metals, chemicals, and other pollutants, posing significant challenges to soil health and agricultural productivity. Soil remediation and restoration technologies offer practical solutions to detoxify and rejuvenate these damaged landscapes. Among the most promising approaches are phytoremediation and bioremediation, both of which leverage natural processes to restore soil health while minimizing ecological disruption.

Phytoremediation is an innovative technique that uses plants to absorb, accumulate, and neutralize harmful substances in the soil. Certain plant species, known as hyperaccumulators, are particularly effective at extracting heavy metals such as lead, cadmium, and arsenic. These plants absorb the pollutants through their roots, storing them in their tissues, which can then be safely harvested and disposed of. Phytoremediation is not only cost-effective but also environmentally friendly, as it enhances soil structure and biodiversity while removing toxins.

Bioremediation takes a microbial approach to soil restoration, deploying bacteria, fungi, or other microorganisms to break down pollutants into less harmful substances. These microbes metabolize contaminants such as hydrocarbons, pesticides, and industrial chemicals, transforming them into harmless byproducts like water and carbon dioxide. Advancements in bioremediation have made it possible to tailor microbial strains to specific pollutants, increasing the efficiency and effectiveness of the process. For instance, genetically modified microbes have been used to target particularly stubborn pollutants, accelerating cleanup efforts.

Case studies from other post-war regions highlight the transformative potential of these techniques. In the Balkans, phytoremediation has been employed to rehabilitate farmland contaminated by heavy metals after years of conflict, enabling the return of agricultural activity. Similarly, bioremediation has been used in Iraq to clean up oil-contaminated soils, restoring ecosystems and supporting local livelihoods.

For Ukraine, these approaches hold immense promise. By integrating phytoremediation and bioremediation into recovery strategies, contaminated lands can be gradually restored to health, supporting agriculture, biodiversity, and sustainable development. These methods not only address the immediate challenge of soil contamination but also lay the groundwork for long-term resilience, ensuring that affected areas can recover and thrive in the years to come. Soil remediation technologies stand as a testament to the power of nature-based solutions, turning environmental crises into opportunities for renewal and growth.

12.3 Renewable Energy for Sustainable Rebuilding

The destruction of energy infrastructure during the Russia-Ukraine war has not only disrupted essential services but also highlighted the fragility of centralized power systems in conflict zones. Rebuilding offers an opportunity to not just restore what was lost but to innovate, replacing damaged infrastructure with renewable energy solutions that are more resilient, sustainable, and environmentally friendly.

Solar farms and wind turbines stand at the forefront of these renewable solutions. Unlike traditional fossil fuel-based systems, these technologies provide clean energy that reduces greenhouse gas emissions and dependency on finite resources. Large-scale solar farms can quickly generate power for urban areas, while wind turbines are particularly suited for rural regions with consistent wind patterns. By diversifying energy sources, these systems reduce vulnerabilities and create a more reliable power supply.

Microgrids and energy storage systems further enhance resilience, offering a decentralized alternative to traditional power grids. Microgrids operate independently or alongside larger grids, ensuring continued operation even during disruptions. Paired with advanced energy storage solutions, such as lithium-ion or solid-state batteries, microgrids can store surplus energy for use during peak demand or outages. This adaptability is especially critical in post-war recovery, where infrastructure may remain unstable for extended periods.

The benefits of transitioning to renewable energy extend beyond immediate functionality. Reducing reliance on fossil fuels not only mitigates environmental damage but also lowers long-term energy costs. Solar and wind systems require minimal maintenance compared to traditional power plants, making them more sustainable investments for regions rebuilding after conflict. Additionally, the integration of renewable technologies can serve as a model for climate-conscious development, showcasing Ukraine's commitment to sustainable recovery.

For Ukraine, the shift to renewable energy represents more than just a practical response to war-torn infrastructure. It is a chance to build a resilient energy network that supports environmental restoration, economic stability, and long-term sustainability. By leveraging renewable technologies, the nation can transform its energy landscape, turning the challenges of destruction into an opportunity for progress and innovation.

Interview with Rajesh Malhotra, Tech Innovator in Disaster Recovery Tools

Rajesh Malhotra is a tech innovator and the founder of ResilientTech, a company specializing in developing cutting-edge tools for disaster recovery and post-conflict reconstruction. His expertise lies in leveraging emerging technologies like drones, AI, and IoT to support recovery efforts in war-affected and disaster-hit regions worldwide.

Q: Rajesh, thank you for joining us. Can you start by explaining how cutting-edge technologies like drones, AI, and IoT are transforming post-war recovery efforts?

A: Thank you for having me. Emerging technologies like drones, artificial intelligence (AI), and the Internet of Things (IoT) have fundamentally changed the way we approach post-war recovery. They allow us to work more efficiently, gather data in real-time, and implement solutions that were unimaginable just a decade ago.

Drones, for example, have become indispensable in assessing damage in conflict zones. They can quickly and safely survey areas that are too dangerous or inaccessible for humans, such as regions with unexploded ordnance or unstable structures. The high-resolution images and thermal data they capture provide invaluable insights for planning reconstruction and environmental restoration.

AI plays a critical role in analyzing the vast amounts of data generated during recovery efforts. From predicting the spread of pollutants in water systems to optimizing supply chains for delivering aid, AI enables faster and more accurate decision-making. Machine learning models can also identify patterns of damage and recommend tailored solutions for specific regions.

IoT technology connects sensors and devices to monitor critical infrastructure in real-time. For example, IoT-enabled water quality sensors can track contamination levels and notify teams of changes that require immediate action. These systems reduce the need for constant manual oversight and improve the efficiency of restoration projects.

Together, these technologies provide an integrated approach to recovery, combining speed, precision, and scalability.

Q: Can you share specific examples of how these technologies have been used in recent post-conflict recovery projects?

A: Absolutely. One notable example is our work in the Donbas region of Ukraine, which has been heavily affected by the conflict. We deployed drones equipped with multispectral cameras to survey damaged industrial sites and map areas contaminated by chemical spills. This data was then fed into an AI platform that modeled the spread of pollutants and recommended priority zones for cleanup.

Another example is a project in Syria, where IoT devices were installed to monitor water distribution systems in areas recovering from infrastructure damage. These sensors detected leaks and contamination in real-time, helping local authorities allocate resources more effectively and restore clean water to affected communities.

We've also used drones for reforestation efforts in post-conflict Rwanda. The drones planted native tree seeds in degraded areas, covering hectares of land much faster than manual efforts would have allowed. AI-powered models then monitored the growth of these trees, ensuring that reforestation was progressing as planned.

These projects demonstrate how technology can address diverse challenges, from environmental restoration to infrastructure rebuilding.

Q: How does international collaboration enhance the deployment of these technologies in post-war recovery?

A: International collaboration is critical for scaling the impact of these technologies. No single organization or country has all the resources or expertise needed for comprehensive recovery efforts, especially in conflict zones where challenges are complex and multifaceted.

For example, in Ukraine, we partnered with international NGOs, local governments, and tech companies to integrate our tools into broader recovery efforts. The data collected by our drones and IoT devices was shared with environmental scientists and urban planners, enabling them to make informed decisions.

Collaboration also facilitates funding. Many of these technologies require significant investment, which can be a barrier for resource-constrained regions. By pooling resources from international organizations like the United Nations and private donors, we've been able to deploy these tools in areas that need them most.

Moreover, knowledge sharing is a key benefit of collaboration. Partnering with experts in various fields—such as hydrology, engineering, and logistics—enhances the effectiveness of our technologies and ensures they're applied in ways that align with local needs.

Q: What challenges do you face in deploying these technologies in conflict zones?

A: Deploying technology in conflict zones comes with significant challenges. First and foremost is **safety**. Many areas are inaccessible due to ongoing violence or the presence of landmines and unexploded ordnance. Even drones, which reduce human risk, can be mistaken for military equipment, making their use in sensitive areas tricky.

Another challenge is **infrastructure**. Many conflict zones lack the basic infrastructure needed to support advanced technologies, such as stable internet connections or electricity. This requires us to develop portable and self-sufficient systems, like drones powered by solar energy or IoT devices with long battery life.

Funding is another issue. While international organizations are increasingly recognizing the value of technology in recovery efforts, securing consistent funding for these tools can be difficult. Donors often prioritize immediate humanitarian needs over long-term recovery technologies, even though the latter can significantly accelerate rebuilding efforts.

Finally, there's the challenge of **local capacity building**. Technology alone isn't enough—it needs skilled operators and maintainers. We invest heavily in training local teams to use and sustain these technologies, ensuring that they remain functional and beneficial long after our deployment ends.

Q: How do you ensure that these technologies are accessible and sustainable for local communities?

A: Accessibility and sustainability are core principles of our work. To ensure accessibility, we focus on developing user-friendly tools that don't require extensive technical expertise. For example, our drones come with intuitive software that local operators can learn to use in just a few days.

We also emphasize **training and capacity building**. In every project, we train local teams to operate, maintain, and repair the technologies we deploy. This not only empowers communities but also ensures the tools remain functional and impactful in the long term.

Sustainability is another priority. We design our technologies to be durable and adaptable to the unique challenges of conflict zones. For instance, we use solar-powered IoT devices that can function without a stable power grid. Additionally, we focus on using materials that are readily available locally to facilitate repairs and replacements.

Finally, we involve communities in the planning and implementation stages of our projects. By aligning our efforts with their priorities and integrating their feedback, we create solutions that are more likely to be accepted and sustained.

Q: What gives you hope as you continue to develop these technologies?

A: What gives me hope is seeing the tangible impact these technologies have on communities. When you witness a village regain access to clean water or see a barren landscape come back to life through reforestation, it's a reminder of what's possible when innovation meets purpose.

I'm also inspired by the growing global commitment to integrating technology into recovery efforts. Governments, NGOs, and private companies are increasingly recognizing the value of these tools, not just for immediate recovery but for building long-term resilience.

Lastly, the people I work with—both my team and the communities we serve—are a constant source of motivation. Their resilience and determination remind me why this work matters and why we must continue pushing the boundaries of what technology can achieve.

Q: What message would you like to share with the global community about the role of technology in post-war recovery?

A: My message is that technology is not a luxury—it's a necessity in post-war recovery. It has the power to save lives, accelerate rebuilding, and create systems that are more resilient to future challenges. However, its success depends on collaboration, investment, and a commitment to ensuring these tools are accessible to those who need them most.

We have the tools to rebuild better and smarter. Now it's up to all of us—governments, organizations, and individuals—to harness this potential and ensure that no community is left behind.

Rajesh Malhotra's insights highlight the transformative potential of technology in post-war recovery. His work demonstrates how innovation, collaboration, and a focus on sustainability can help rebuild not just infrastructure, but hope and resilience in the face of devastation.

12.4 Water Purification and Resource Management

The war in Ukraine has left many of its rivers and wetlands heavily polluted by chemicals and debris, posing significant risks to public health, agriculture, and biodiversity. Addressing these challenges requires innovative approaches to water purification and resource management, combining advanced technologies with nature-based solutions to ensure long-term resilience and sustainability.

Advanced filtration systems are essential for treating chemically polluted water sources. These systems use state-of-the-art technologies such as reverse osmosis, activated carbon filters, and nanotechnology to remove harmful contaminants, including heavy metals, pesticides, and industrial chemicals. Portable filtration units can provide immediate relief in areas with limited access to clean water, while larger systems can be deployed to restore major rivers and wetlands affected by the conflict.

Constructed wetlands offer a complementary, nature-based solution to water purification. These engineered ecosystems mimic natural wetland processes, using vegetation, soil, and microorganisms to filter pollutants from water. Constructed wetlands not only improve water quality but also create habitats for wildlife, support biodiversity, and enhance the aesthetic and ecological value of the landscape. They are particularly suited to regions with degraded ecosystems, offering a cost-effective and sustainable alternative to conventional water treatment methods.

The integration of IoT-enabled water management systems further enhances efficiency and effectiveness. IoT sensors can monitor water quality in real time, measuring parameters such as pH, turbidity, and pollutant levels. These sensors provide data that can be analyzed to identify pollution hotspots, track the progress of purification efforts, and optimize resource allocation. IoT systems also enable smart irrigation practices, ensuring that water is used efficiently in agriculture and reducing waste in regions recovering from resource shortages.

By combining advanced filtration technologies, nature-based solutions, and IoT-enabled management, Ukraine can address the immediate challenges of water pollution while laying the foundation for sustainable water resource management. These approaches not only restore access to clean water but also build resilience against future disruptions, ensuring that water remains a vital and protected resource for communities and ecosystems alike. In the context of post-war recovery, the role of water purification and management is pivotal in supporting health, agriculture, and biodiversity, offering a path toward a cleaner and more sustainable future.

12.5 Reforestation and Habitat Restoration Innovations

The war in Ukraine has caused significant damage to forests and habitats, disrupting ecosystems and threatening biodiversity. Restoring these vital natural resources requires innovative approaches that leverage advanced technologies and community engagement. Reforestation and habitat restoration innovations, including drone technology, genetic tools, and citizen science platforms, offer scalable and efficient solutions to address the widespread environmental devastation.

Drones are revolutionizing reforestation efforts by enabling large-scale seed dispersal over difficult-to-reach or heavily damaged areas. Equipped with advanced mapping systems, drones can identify optimal planting locations and distribute seeds efficiently, accelerating the reforestation process. This technology not only reduces the labor and time required for traditional replanting methods but also ensures greater precision in re-establishing forest cover. In Ukraine, drones can play a pivotal role in restoring forests damaged by bombings or illegal logging during the conflict.

Genetic tools offer another cutting-edge approach to rebuilding biodiversity hotspots and restoring native species. Techniques such as genome editing and seed banking allow conservationists to reintroduce resilient plant varieties tailored to local environmental conditions. These tools can help combat the spread of invasive species, enhance the survival rates of reintroduced flora, and support the recovery of ecosystems critical to wildlife and community well-being.

Citizen science platforms further amplify the impact of restoration efforts by involving local communities in monitoring and maintaining restored habitats. Through mobile apps and online platforms, citizens can report changes in forest health, track wildlife activity, and contribute to data collection. This participatory approach fosters a sense of stewardship among residents, empowering them to play an active role in preserving their environment. It also provides valuable, on-the-ground insights to complement scientific research and inform adaptive management strategies.

By integrating drones for rapid reforestation, genetic tools for biodiversity recovery, and citizen science for community involvement, Ukraine can rebuild its natural ecosystems with innovation and inclusivity. These solutions not only address the immediate need for environmental restoration but also lay the groundwork for sustainable habitat management in the future. Reforestation and habitat restoration efforts are essential to healing the ecological scars of war, ensuring that forests and ecosystems once again thrive as pillars of biodiversity and resilience.

12.6 Technology for Climate Resilience

As Ukraine rebuilds in the aftermath of war, incorporating climate resilience into recovery efforts is critical to ensuring long-term sustainability. Advanced technologies offer innovative solutions to address the dual challenges of reconstruction and climate change. From AI models that simulate climate impacts to green building technologies and tools for mitigating greenhouse gas emissions, these advancements provide a framework for creating infrastructure and urban spaces that are both adaptive and sustainable.

AI models are at the forefront of climate resilience planning. These tools use machine learning and predictive analytics to simulate the effects of climate change on infrastructure and ecosystems. By analyzing historical data and current trends, AI can forecast risks such as flooding, heatwaves, or rising sea levels, enabling planners to design adaptive infrastructure that withstands future challenges. For instance, AI can inform the placement and construction of flood-resistant roads, energy grids, and water management systems, ensuring that rebuilt infrastructure remains functional in the face of changing environmental conditions.

Green building technologies are transforming urban spaces by prioritizing energy efficiency and sustainability. Innovations such as passive solar design, energy-efficient materials, and renewable energy systems reduce the carbon footprint of construction while enhancing the resilience of buildings to extreme weather events. These technologies also incorporate natural ventilation, insulation, and water-saving systems to minimize resource consumption. In post-war Ukraine, green building approaches can help create urban environments that are both sustainable and equipped to cope with future climate risks.

Tools for monitoring and mitigating greenhouse gas emissions play a vital role during reconstruction. IoT-enabled sensors and satellite-based monitoring systems can track emissions from construction activities and industrial processes, providing real-time data to ensure compliance with environmental standards. Carbon offset technologies, such as reforestation projects and carbon capture systems, can further reduce the impact of rebuilding efforts on the environment. By integrating these tools into the recovery process, Ukraine can align its reconstruction goals with global climate commitments, contributing to broader efforts to combat climate change.

By leveraging AI for predictive modeling, adopting green building practices, and employing advanced emission monitoring tools, Ukraine can transform its recovery efforts into a model of climate resilience. These technologies not only address the immediate challenges of rebuilding but also position the nation as a leader in sustainable development. Incorporating climate resilience into post-war reconstruction ensures that the infrastructure and communities rebuilt today will endure and thrive in the face of future environmental challenges.

12.7 International Collaborations and Funding for Technological Solutions

The scale of environmental and infrastructural recovery required after the Russia-Ukraine war necessitates not just national efforts but also robust international collaborations. Global organizations, governments, and private-sector entities play a pivotal role in providing the technical expertise, financial resources, and innovative technologies essential for effective recovery. These partnerships have the potential to transform the rebuilding process into a model of sustainable and resilient development.

Global organizations, such as the United Nations, World Bank, and European Union, have a long history of supporting post-conflict recovery efforts. Their involvement in Ukraine includes technical assistance in deploying advanced technologies, such as AI-powered environmental monitoring and renewable energy systems, as well as financial aid for large-scale restoration projects. For instance, the Green Climate Fund could provide critical support for implementing sustainable water management systems and renewable energy infrastructure, ensuring that rebuilding efforts align with global climate goals.

Examples of successful international partnerships offer a blueprint for Ukraine's recovery. In Iraq, collaborations between local governments, NGOs, and international bodies enabled the deployment of IoT-based water monitoring systems to restore essential resources. Similarly, in post-war Rwanda, international partnerships facilitated reforestation projects using drone technology, revitalizing ecosystems and supporting local livelihoods. These case studies demonstrate the importance of cross-border collaboration in harnessing cutting-edge technologies for environmental restoration and sustainable development.

To ensure the long-term sustainability of these efforts, robust frameworks must be established. These frameworks should prioritize technology-driven initiatives that integrate environmental, social, and economic goals. For example, public-private partnerships can drive innovation while ensuring accountability and scalability. International agreements and standards can help maintain transparency, encourage resource-sharing, and align recovery efforts with global sustainability benchmarks.

For Ukraine, international collaboration offers not just resources but also opportunities to share knowledge, leverage expertise, and build a resilient future. By engaging with global partners, the nation can access advanced technologies and financial mechanisms that enable comprehensive recovery while fostering long-term sustainability. These partnerships are essential for transforming the immense challenges of post-war rebuilding into an opportunity for progress, innovation, and resilience on both a national and global scale.

Conclusion: Technology as a Pillar of Resilient Recovery

The transformative potential of technology in post-war recovery is undeniable. From advanced environmental monitoring tools to innovative approaches in soil remediation, renewable energy, and water management, technology offers Ukraine the means to not only rebuild but to emerge stronger and more resilient. These solutions address immediate challenges while laying the foundation for sustainable development, ensuring that recovery efforts are aligned with both national priorities and global environmental goals.

Strategic investments in cutting-edge technologies are essential to harness their full potential. The integration of AI, IoT, and renewable systems requires robust infrastructure, skilled expertise, and a commitment to long-term sustainability. International collaborations will play a critical role in providing the technical and financial support needed to scale these innovations. By forging partnerships with global organizations, private-sector entities, and other nations, Ukraine can access the resources and knowledge necessary for comprehensive recovery.

Ukraine stands at a pivotal moment in its history—a moment that presents not only immense challenges but also unprecedented opportunities. By leveraging technology as a cornerstone of its recovery strategy, the nation has the potential to become a global leader in sustainable reconstruction. Ukraine's journey can serve as a model for integrating resilience, innovation, and sustainability into post-conflict rebuilding, inspiring other nations facing similar trials.

As the scars of war begin to heal, technology offers a pathway to a brighter and more secure future. With strategic investments, international collaboration, and a forward-thinking vision, Ukraine can transform its recovery into a testament to human ingenuity and resilience. In doing so, it will not only rebuild its own infrastructure and ecosystems but also contribute to a global movement toward sustainable and resilient development.

Conclusion: A War Beyond Borders

The Russia-Ukraine conflict has underscored the far-reaching consequences of modern warfare, extending beyond traditional battlefields to the interconnected realms of the environment and cyberspace. Its impacts have been global, revealing vulnerabilities in critical infrastructure, ecosystems, and digital systems, while challenging international norms and cooperation. As the world grapples with the cascading effects of this conflict, it becomes increasingly clear that action is needed to prevent further collateral damage and build a more resilient future. Despite the destruction, there is hope—rooted in the potential for international unity and innovation to drive recovery and prevention. This conclusion reflects on the global implications of this war, the urgency for proactive measures, and the enduring hope for a more secure and sustainable world.

A War Beyond Borders: The Global Implications of the Russia-Ukraine Conflict on the Environment and Cyberspace

The Russia-Ukraine conflict has reverberated far beyond the borders of the warring nations, leaving indelible marks on ecosystems, global resource stability, and the increasingly interconnected realm of cyberspace. This is not merely a regional war—it is a harbinger of how modern conflicts transcend geographical boundaries, spilling over into domains that impact the entire world. As the shockwaves of this war ripple outward, they reveal vulnerabilities and interconnectedness that demand urgent global attention.

Environmental Impact Across Borders

The environmental consequences of this conflict have reached far beyond the immediate theaters of war, cascading into broader ecological and resource-related crises. Ukraine, known as one of Europe's most fertile agricultural regions and a significant energy transit hub, has seen its environmental stability shaken, with impacts that extend to the global stage.

Energy Disruptions are one of the most immediate examples of these far-reaching effects. Cyberattacks targeting pipelines and energy grids have not only disrupted local supplies but have also exacerbated global energy insecurity. These disruptions have forced nations to lean on less sustainable energy alternatives, such as coal or diesel, increasing emissions and compounding the global climate crisis. In this way, a cyber breach in one region can directly impact international efforts to transition to greener energy systems.

The war has also resulted in **transboundary pollution**, as attacks on industrial facilities have released pollutants into the air, water, and soil. For instance, toxic chemicals leaking from bombed factories and storage facilities have traveled via waterways and atmospheric currents, threatening ecosystems far from the conflict zone. Neighboring countries, already burdened with supporting displaced populations, now contend with the environmental fallout as well.

Agricultural shocks have further demonstrated the interconnectedness of global systems. Ukraine, often referred to as the "breadbasket of Europe," plays a vital role in feeding the world. The disruption of its agricultural exports has not only led to food shortages in importing nations but has also strained local ecosystems in those regions. Increased agricultural production to fill the gap has intensified soil degradation and water resource depletion, creating a chain reaction that illustrates the fragility of global food systems.

Cyber Impacts with Global Reach

The digital dimension of the Russia-Ukraine conflict has highlighted the boundless nature of cyber warfare. What begins as a targeted cyberattack on Ukrainian systems can rapidly escalate into a global incident, revealing the vulnerabilities inherent in an interconnected digital world.

Cross-border malware spread is a prime example. The NotPetya malware attack, initially aimed at disrupting Ukrainian systems, quickly spiraled out of control, infecting multinational corporations and critical infrastructure worldwide. Companies far removed from the conflict zone suffered billions in damages, showcasing how localized cyber incidents can ripple across the global economy. These events underscore the fragility of digital networks and the urgent need for international cooperation in securing them.

Beyond the technical disruptions, the war has also **undermined trust in cyberspace**. The proliferation of misinformation and psychological operations (PsyOps) has polarized societies, deepened divisions, and eroded trust in governments and institutions. These tactics, amplified through social media, have not only influenced perceptions of the conflict but have also left long-lasting scars on public discourse globally. The destabilization of trust is a slow-burning consequence of cyber warfare that will take years to repair.

The war has further exposed **global cybersecurity challenges**, emphasizing how unprepared the world is for large-scale, state-sponsored cyberattacks. The absence of robust international frameworks and norms to manage such threats has left nations and organizations scrambling to respond, often with inconsistent or inadequate measures. This gap in preparedness underscores the critical need for coordinated international responses and stronger cybersecurity policies to protect against evolving threats.

A Precedent for Hybrid Warfare

Perhaps the most significant implication of the Russia-Ukraine conflict is the precedent it sets for **hybrid warfare**—a multidimensional strategy that combines traditional military tactics with cyber operations and environmental impacts. By blurring the lines between military and civilian targets, hybrid warfare amplifies the stakes and complicates the responses of adversaries.

The integration of cyberattacks with physical warfare, such as disabling energy systems during a ground offensive, has shown how these strategies can magnify their effectiveness. Simultaneously, attacks on eco-sensitive infrastructure, like water systems or industrial plants, serve both as tactical maneuvers and psychological operations, creating fear and instability beyond the battlefield.

Hybrid warfare has also highlighted the vulnerability of civilian infrastructure, transforming previously untouchable sectors—such as hospitals, energy grids, and water management systems—into targets. The ripple effects of these strategies extend well beyond immediate tactical gains, disrupting global systems and threatening long-term stability.

This integration of cyber and environmental dimensions sets a dangerous precedent for future conflicts. As more nations develop hybrid warfare capabilities, the risks to ecosystems, civilian infrastructure, and digital networks grow exponentially. The lessons from this war must be used to strengthen global resilience, ensuring that future conflicts do not wreak similar havoc across interconnected domains.

In short, the Russia-Ukraine conflict has demonstrated that modern warfare is no longer confined to battlefields; its effects are felt in every corner of the globe. The environmental and cyber dimensions of this war underscore the urgency for international collaboration, robust governance, and innovative solutions to address the challenges of a world where borders no longer constrain the impacts of conflict. This is a war beyond borders—one that demands global action to protect the systems and ecosystems that sustain us all.

The Urgency for Action: Preventing Future Collateral Damage in Both Realms

The Russia-Ukraine conflict has made one truth abundantly clear: the interconnected challenges of cyber warfare and environmental security demand immediate and decisive action. As the fallout from this war ripples across ecosystems and digital networks, it is evident that the systems we depend on—both physical and digital—are more vulnerable than ever. The path forward requires resilience, regulation, collaboration, and education, as well as a shared commitment to safeguarding the delicate balance between technological progress and environmental sustainability.

Strengthening Cyber and Environmental Resilience

The first and most pressing step is to build resilience in both cyberspace and eco-sensitive systems. The lessons from this conflict underscore the need for robust defenses capable of mitigating threats and withstanding disruptions.

In cyberspace, resilience begins with investments in advanced cybersecurity measures. Governments and organizations must embrace tools like AI-driven threat detection systems that can analyze patterns, predict potential breaches, and neutralize threats before they escalate. Cross-border intelligence sharing will be crucial to counter sophisticated cyberattacks that target interconnected global systems. Just as cyber threats know no borders, so too must our defenses transcend national boundaries, fostering a collaborative approach to safeguarding critical networks.

For the environment, resilience means redesigning infrastructure to withstand both physical and cyber threats. Real-time monitoring systems and automated fail-safes should be standard in eco-sensitive infrastructure, ensuring that potential disruptions—from cyberattacks on dams to industrial sabotage—can be detected and mitigated before they lead to ecological disasters. These measures not only protect against immediate risks but also ensure that recovery from such events is faster and more sustainable.

Developing Global Norms and Regulations

The lack of clear international norms governing cyber and environmental warfare has left a dangerous void, allowing unchecked escalation and unintended consequences. Addressing this gap is critical to preventing future harm.

In cyberspace, global norms must prohibit attacks on civilian infrastructure and eco-sensitive systems. These agreements should explicitly safeguard critical resources like water treatment facilities, energy grids, and agricultural systems, which are essential for both human survival and ecological balance. Establishing these protections will deter bad actors and create a framework for accountability when violations occur.

For environmental impacts, frameworks must address the long-term consequences of conflicts. Holding perpetrators accountable for environmentally harmful actions—such as polluting rivers, destroying forests, or targeting industrial facilities—will not only deter violations but also encourage sustainable recovery practices. These frameworks should include mechanisms for reparations and support for affected regions, ensuring that the burden of recovery does not fall solely on local communities.

The creation and enforcement of such norms will require multilateral cooperation and the political will to prioritize global stability over short-term national interests. By fostering a sense of shared responsibility, the international community can take meaningful steps toward mitigating the dual threats of cyber and environmental warfare.

Fostering International Collaboration

The interconnected nature of these challenges demands a united global response. No single nation, organization, or sector can address the complexities of cyber and environmental threats alone.

Multilateral organizations such as the United Nations, NATO, and regional alliances must take the lead in coordinating efforts to address these dual challenges. Their role is not only to facilitate dialogue but also to implement actionable solutions. This includes establishing cross-border partnerships, sharing resources, and creating joint response mechanisms for cyber and environmental emergencies. For example, a unified approach to protecting shared water resources or transboundary energy systems could prevent cascading failures caused by localized attacks.

Public-private partnerships are equally essential. The private sector holds much of the expertise and innovation needed to address these threats, while governments provide the regulatory frameworks and oversight necessary for equitable deployment. By working together, these entities can bridge the gap between innovation and implementation, ensuring that advanced technologies—such as IoT-enabled environmental monitors or AI-driven cybersecurity platforms—are used effectively and equitably.

Investing in Education and Capacity Building

Long-term resilience requires not just technology but also the human expertise to manage it. Building awareness and technical capacity across sectors is critical for reducing vulnerabilities and fostering a new generation of leaders equipped to address these intertwined challenges.

Cybersecurity education must go beyond technical skills to emphasize the environmental implications of digital threats. Professionals in this field need to understand how cyberattacks can disrupt ecosystems, harm biodiversity, and undermine sustainability efforts. Cross-disciplinary programs that integrate environmental science with cybersecurity will be essential for creating a workforce capable of addressing these complex risks.

Training for environmental managers and policymakers is equally important. These stakeholders must understand the potential digital vulnerabilities of eco-sensitive systems, from water treatment plants to renewable energy grids. By equipping them with the knowledge to assess and mitigate digital risks, we can ensure more informed decision-making at all levels.

Educational efforts should also extend to the broader public. Raising awareness about the interconnected nature of cyber and environmental challenges will foster a culture of vigilance and shared responsibility, empowering communities to play an active role in protecting the systems they rely on.

Thus, the urgency for action is undeniable. The Russia-Ukraine conflict has shown us how deeply intertwined our digital and natural worlds are—and how vulnerable they can be without proactive measures. By strengthening resilience, establishing global norms, fostering collaboration, and investing in education, the international community can prevent future collateral damage in both realms. The time to act is now, before the lessons of this conflict are overshadowed by the next crisis. Together, we have the tools, knowledge, and resolve to build a more secure and sustainable future for all.

Hope Amid Destruction: The Role of International Unity in Recovery and Prevention

Amid the devastation caused by the Russia-Ukraine conflict, glimmers of hope have emerged from the collective efforts of nations, organizations, and communities to confront its far-reaching impacts. These responses underscore the transformative power of international collaboration in driving recovery and fostering resilience against future challenges. While the scars of this war remain, the unity and innovation demonstrated in its aftermath offer a blueprint for a more secure and sustainable world.

Learning from the Lessons of Unity

The global response to the Russia-Ukraine conflict has highlighted the immense potential of collective action. Despite the challenges, nations and organizations have joined forces to support Ukraine's recovery efforts, demonstrating that even in times of crisis, collaboration can prevail.

Nations and institutions across the world have provided funding, technical expertise, and advanced tools to bolster Ukraine's cyber and environmental defenses. From safeguarding critical infrastructure against cyber threats to initiating ecological recovery projects, this support reflects a shared commitment to addressing the multidimensional impacts of modern warfare. Platforms like the European Union's Cybersecurity Strategy and the United Nations' environmental recovery programs have become models of coordinated action, illustrating how global partnerships can mitigate the consequences of even the most complex crises.

These collaborative efforts go beyond addressing immediate needs. They represent a collective acknowledgment of the interconnected nature of today's challenges, setting a precedent for tackling future conflicts through unity and shared responsibility.

Innovative Solutions for a Resilient Future

Amid the destruction, technological advancements have emerged as beacons of hope, offering transformative solutions for building resilience and preventing future harm. The tools of the digital age—when deployed responsibly—can mitigate the impacts of war and drive sustainable recovery.

AI and satellite imagery are revolutionizing how risks are identified and managed. These technologies enable real-time monitoring of conflict zones, providing actionable insights into ecological and cyber vulnerabilities. Predictive analytics powered by AI can foresee emerging threats, allowing nations to respond proactively rather than reactively.

IoT and smart systems have proven invaluable in enhancing the resilience of critical infrastructure. These interconnected networks ensure continued operation during disruptions, protecting essential services such as energy grids, water systems, and healthcare facilities. By integrating IoT devices with advanced analytics, nations can strengthen their defenses against cascading failures.

Sustainable technologies further highlight the potential for innovation to reduce the environmental footprint of recovery efforts. Renewable energy systems, green construction materials, and eco-friendly rebuilding practices can transform devastated regions into models of sustainability, ensuring that recovery efforts align with long-term environmental goals.

These technological advancements not only address the immediate consequences of conflict but also create opportunities to reimagine and rebuild systems in ways that are more sustainable, equitable, and prepared for future challenges.

A Vision for Global Solidarity

The interconnected nature of cyber and environmental challenges underscores the importance of global solidarity in addressing shared risks. The impacts of the Russia-Ukraine conflict have shown that no nation can face these threats alone. Instead, collaborative efforts rooted in mutual support and shared goals must guide the path forward.

Global solidarity requires nations to move beyond short-term interests and prioritize long-term stability and sustainability. This means recognizing that the actions of one country can have far-reaching consequences, and that cooperation, not competition, is the key to navigating the complexities of the modern world.

Collaborative efforts, such as international agreements on cybersecurity norms and environmental protections, strengthen bonds between nations and build trust in the face of uncertainty. By working together to address shared risks, the global community can create a more secure and cooperative future, where progress is achieved through unity rather than division.

Rebuilding Better and Stronger

Recovery efforts in the wake of the Russia-Ukraine conflict must go beyond simply restoring what was lost. This moment presents an opportunity to build systems that are more resilient, sustainable, and inclusive, laying the groundwork for a secure and prosperous future.

Post-war reconstruction provides a unique chance to integrate advanced technologies and eco-friendly practices into rebuilding efforts. Smart grids powered by renewable energy, green infrastructure that prioritizes sustainability, and digital tools that enhance resilience can transform devastated regions into beacons of progress. These innovations not only address immediate needs but also ensure that rebuilt systems are better equipped to withstand future challenges.

By addressing both environmental and cyber vulnerabilities, nations can ensure that recovery efforts contribute to global stability and resilience. This holistic approach to rebuilding—one that prioritizes sustainability, equity, and preparedness—creates a foundation for enduring peace and prosperity.

A Hopeful Path Forward

While the Russia-Ukraine conflict has brought immense destruction, it has also revealed the strength and potential of international unity. The lessons of this war remind us that even in the darkest moments, collaboration and innovation can light the way toward recovery. By leveraging technological advancements, fostering global solidarity, and prioritizing sustainable practices, the world can transform the challenges of today into opportunities for a brighter future.

This vision of unity, resilience, and innovation serves as a testament to what is possible when nations come together to address shared risks. In the face of destruction, hope endures—not as a passive ideal, but as an active force driving recovery, prevention, and progress for generations to come.

A Path Forward

The Russia-Ukraine conflict has revealed the profound interconnection between cyberspace and the environment, emphasizing the need for integrated solutions to address these dual challenges. While the damage has been extensive, it has also catalyzed global awareness and innovation, providing a roadmap for navigating the complexities of modern warfare.

The urgency for action cannot be overstated. Preventing future collateral damage requires resilience, collaboration, and a commitment to sustainability. By leveraging technology, fostering international unity, and prioritizing ethical responsibility, the global community can transform the lessons of this conflict into a blueprint for a more secure and sustainable world. Amid the destruction, hope endures—rooted in the collective resolve to rebuild stronger, protect ecosystems, and secure the digital and natural realms for generations to come.

Appendices

The appendices provide additional resources and context to enhance understanding of the topics covered in this book. They include a glossary of key terms, a timeline of major events, and a curated list of recommended readings and resources for further exploration.

Glossary of Terms

This glossary defines key terms related to cybersecurity and environmental issues discussed in the book, offering readers a quick reference for technical and specialized concepts.

Artificial Intelligence (AI): A branch of computer science that enables machines to perform tasks typically requiring human intelligence, such as learning, reasoning, and decision-making.

Cyberattack: An attempt to gain unauthorized access to a computer system, network, or device with the intent to disrupt, damage, or steal data.

Deforestation: The clearing or thinning of forests, often resulting from human activity, which can lead to habitat loss, climate change, and biodiversity decline.

Internet of Things (IoT): A network of physical devices connected via the internet, enabling them to collect, share, and act on data in real time.

Intrusion Detection System (IDS): A cybersecurity tool designed to monitor network activity and identify potential unauthorized access or malicious actions.

Phytoremediation: A process that uses plants to remove, transfer, or stabilize contaminants in soil, water, or air.

Renewable Energy: Energy derived from natural sources that are replenished over time, such as sunlight, wind, and water.

Satellite Imagery: High-resolution images captured by satellites used for monitoring environmental conditions, assessing damage, and planning recovery efforts.

Smart Grids: Electricity networks that use digital technology to optimize the production, distribution, and consumption of energy.

Sustainable Development: Development that meets the needs of the present without compromising the ability of future generations to meet their own needs.

Timeline of Major Events: Environmental and Cyber Impacts

This timeline highlights key incidents and developments related to environmental and cyber impacts in conflict zones, providing historical context for the themes explored in the book.

2015-2016: Ukraine Power Grid Attacks

A series of cyberattacks targeted Ukraine's power grid, causing widespread blackouts and exposing vulnerabilities in critical infrastructure.

2017: NotPetya Malware Attack

Initially aimed at Ukrainian systems, this cyberattack spread globally, affecting multinational corporations and highlighting the risks of cross-border cyber threats.

2021: Colonial Pipeline Ransomware Attack

A cyberattack disrupted the largest fuel pipeline in the United States, leading to regional fuel shortages and increased emissions from alternative transportation methods.

2021: Oldsmar Water Treatment Plant Cyberattack

An attempted cyberattack targeted a Florida water treatment plant, aiming to manipulate chemical levels and potentially endanger public health.

2022: Russia-Ukraine War

Environmental damage and cyberattacks escalated during the conflict, including deforestation, oil spills, and disruptions to energy grids, emphasizing the need for integrated recovery strategies.

Ongoing: Global Climate and Cybersecurity Initiatives

Efforts like the Paris Agreement and international cybersecurity frameworks continue to address the interconnected challenges of environmental sustainability and digital resilience.

Recommended Reading and Resources

This curated list includes books, reports, and online resources for readers interested in further exploring the intersection of environmental and cyber issues.

Books:

Cybersecurity and Cyberwar: What Everyone Needs to Know by P.W. Singer and Allan Friedman

An accessible introduction to cybersecurity, addressing its relevance in modern conflicts.

This Changes Everything: Capitalism vs. The Climate by Naomi Klein

A thought-provoking exploration of how climate change intersects with global economic and political systems.

Earth Under Siege: From Air Pollution to Global Change by Richard Turco

A detailed account of environmental challenges and the technologies used to address them.

Reports:

Global Risks Report (World Economic Forum) An annual report analyzing key risks to global stability, including cyber threats and environmental degradation.

Environmental Dimensions of Security (United Nations Environment Programme)

A report examining how environmental issues influence global security dynamics.

Web Resources:

European Space Agency's Copernicus Programme: https://www.copernicus.eu

A comprehensive platform for satellite-based environmental monitoring.

National Institute of Standards and Technology (NIST) Cybersecurity Framework: https://www.nist.gov/cyberframework

Guidelines for improving cybersecurity practices across industries.
World Resources Institute (WRI): https://www.wri.org
A leading organization focused on environmental research and policy.

Acknowledgment

The completion of this book would not have been possible without the guidance, support, and contributions of numerous individuals and organizations who shared their expertise, insights, and encouragement throughout this journey.

First and foremost, I am deeply grateful to the experts, researchers, and practitioners who generously shared their knowledge, experiences, and perspectives. Their contributions provided invaluable depth to the discussions on the intersection of cyber warfare and environmental challenges.

I extend my heartfelt thanks to the environmental scientists, cybersecurity analysts, and technology innovators whose work continues to inspire and inform solutions for a more secure and sustainable future. Their dedication to addressing the complex challenges of our time serves as a beacon of hope.

A special acknowledgment goes to my family and friends for their unwavering support and patience during the writing process. Your encouragement and understanding have been my greatest source of strength.

I also wish to thank the publishers, editors, and reviewers who guided this project from concept to completion. Your expertise and feedback helped shape the book into its final form, ensuring clarity and coherence in presenting these critical issues.

Lastly, to the readers—your curiosity and commitment to understanding the interplay between technology, conflict, and the environment are what make these conversations meaningful. I hope this book sparks new ideas, discussions, and actions that contribute to building a resilient and equitable world.

Thank you all for being part of this journey. This work is as much yours as it is mine. Together, we can rise to the challenges of today and create a brighter tomorrow.

About the Series

Echoes of War: The Russia-Ukraine Conflict and Its Global Ripples by Sandeep Chavan

Exploring the profound impact of the Russia-Ukraine war across humanity, geopolitics, and beyond.

The Russia-Ukraine war, which began in 2022, has left an indelible mark on the world. It is a conflict that transcends borders, reshaping not just physical territories but also the political, economic, and environmental landscapes. *Echoes of War: The Russia-Ukraine Conflict and Its Global Ripples* is a compelling book series that examines the wide-reaching consequences of this historic conflict, offering nuanced insights into its various dimensions.

Titles within the Series

Humanitarian Impact

Shattered Lives: The Human Cost of the Russia-Ukraine War

Displacement, Resilience, and Hope Amid Conflict

A deep exploration of the human toll of war, focusing on displacement, refugee crises, and the enduring resilience of those affected.

Economic Effects

The War Economy: Global Disruptions from the Russia-Ukraine Crisis

How Conflict Redefined Trade, Energy, and Food Security

Analyzing the economic upheaval caused by the conflict, this book explores shifts in global trade, energy markets, and food security.

Geopolitical Shifts

Borders Redrawn: Global Politics in the Wake of the Russia-Ukraine War

NATO, Alliances, and the Changing Face of Power

A detailed look at how the war has reshaped alliances, redrawn geopolitical borders, and transformed global power dynamics.

Military Dynamics

Arms and Allies: Military Strategies of the Russia-Ukraine War

Tactics, Technology, and the Global Arms Race

This book dives into the military strategies, advanced technologies, and evolving alliances that have defined the conflict.

Environmental and Cyber Impacts

Collateral Damage: The Russia-Ukraine War's Toll on Earth and Cyberspace

Environmental Ruin and the Rise of Cyber Warfare

Exploring the hidden costs of war, this volume examines environmental devastation and the escalating role of cyber warfare.

Global Awareness and Legal Challenges

Justice in the Ashes: Accountability in the Russia-Ukraine War

War Crimes and the Pursuit of International Law

A compelling analysis of war crimes, international law, and the global pursuit of accountability and justice.

Discover the Series

Through comprehensive research, compelling narratives, and a global perspective, *Echoes of War* provides a definitive account of how the Russia-Ukraine conflict continues to reshape our world. Each book in the series examines a distinct aspect of the war's far-reaching consequences, offering readers a deeper understanding of its ripple effects across humanity, nations, and the planet.

Explore the *Echoes of War* series and join the conversation on one of the most defining conflicts of our time.

About the Author

Sandeep Chavan is an accomplished author, educator, and thinker with a deep interest in geopolitics, global economics, and their far-reaching impact on humanity. With decades of experience in industrial engineering and education, Chavan brings a unique blend of analytical rigor and empathetic understanding to his work. His writing is characterized by a keen ability to connect complex global phenomena with their real-world implications, making intricate subjects accessible to a wide audience.

In the *Echoes of War* series, Chavan explores the multidimensional impact of the Russia-Ukraine conflict on global politics, economies, societies, and ecosystems. Driven by an unrelenting curiosity to understand the ripple effects of war, he examines not just the immediate consequences but also the deeper, systemic changes that are reshaping the global order. His work reflects a commitment to uncovering truths, highlighting resilience, and fostering dialogue about the challenges and opportunities of an interconnected world.

Chavan's background in industrial engineering provides him with a systems-thinking approach, enabling him to analyze how geopolitical shifts and economic disruptions interconnect and cascade across sectors. His expertise is evident in his ability to dissect the complexities of international trade, energy markets, and technological advancements, and to contextualize these within the broader geopolitical framework.

As an author, Chavan is deeply committed to shedding light on the human cost of conflict while exploring pathways for resilience and recovery. His insights are informed by extensive research, global case studies, and a profound empathy for those whose lives are shaped by the forces of geopolitics. Through the *Echoes of War* series, Chavan not only documents the transformative impacts of the Russia-Ukraine war but also inspires readers to engage critically with the forces shaping our world today.

Chavan's work goes beyond analysis—it is a call to action, urging readers to consider their role in fostering a more equitable and sustainable global order. Whether examining humanitarian crises, economic disruptions, or the evolving dynamics of power, Chavan's voice is a guiding light in understanding the complexities of a world in flux.

Don't miss out!

Visit the website below and you can sign up to receive emails whenever SANDEEP CHAVAN publishes a new book. There's no charge and no obligation.

https://books2read.com/r/B-A-EPGPC-IEJLF

BOOKS 2 READ

Connecting independent readers to independent writers.

Did you love *Collateral Damage*? Then you should read *The Triangular Dynamics*[1] by SANDEEP CHAVAN!

[2]

In an era defined by geopolitical complexities and shifting global power dynamics, The Triangular Dynamics: India between China and USA delves into the intricate relationships shaping the 21st-century world order. Written by Sandeep Chavan, this insightful book explores India's pivotal role as a mediator, balancer, and emerging global leader amidst the superpower rivalry between China and the United States.The book examines how India navigates its relationships with two of the most influential nations in the world. By analyzing India's strategic autonomy, diplomatic engagements, and economic strategies, Chavan sheds light on the country's ability to balance its partnerships with the U.S. and China while advancing its own national interests. This balancing act is not merely a survival tactic but a reflection of India's growing stature in global geopolitics.Drawing on a comprehensive analysis of trade, technology, and diplomacy, The Triangular Dynamics

1. https://books2read.com/u/47qPpq
2. https://books2read.com/u/47qPpq

highlights India's initiatives in reshaping global supply chains, fostering innovation, and championing sustainability. From the Production-Linked Incentive (PLI) schemes to the International Solar Alliance, the book showcases India's efforts to redefine its economic and environmental responsibilities on the global stage.Chavan also explores India's leadership in multilateral platforms like BRICS, the Quad, and the G20, positioning it as a voice for the Global South. The book discusses India's role in addressing critical issues such as climate change, digital governance, and public health, emphasizing its ability to act as a bridge between developed and developing nations.With a focus on India's strategic location, rich cultural legacy, and ambitious vision for 2030, The Triangular Dynamics outlines a roadmap for the nation to emerge as a key player in a multipolar world. It underscores the importance of resilience, innovation, and inclusive growth in shaping a more equitable and harmonious global order.Ideal for readers of international relations, political science, and global economics, this book provides a balanced perspective on the challenges and opportunities within the U.S.-China-India triangle. The Triangular Dynamics: India between China and USA is a thought-provoking exploration of the strategies and values that will shape India's role in an interconnected and competitive world.

Read more at https://www.sandeepjchavan.com/.

Also by SANDEEP CHAVAN

Echoes of War: The Russia-Ukraine Conflict and Its Global Ripples
Borders Redrawn
Arms and Allies
Collateral Damage
Shattered Lives
The War Economy

Ultra Modern Life Philosophy
The Duality of Laughter and Tears
Compare to Accept

Standalone
It's Not AI, It's AHI - Amplified Human Intelligence
The Decision Paradox
The Four Sapiens
Malicious Script of Indian Polity
The IIT Legacy & Global Impact
The IIT Legacy & Global Impact
You are just a Version of Your Original
Win the Game You Didn't Choose
Always Keep Your Bags Packed
Reveal to Shield - Learn the Game of Facades

Engage Beyond Elections
Culture, Identity & Change: The Evolution of Indian Society
The Triangular Dynamics
Never Trust Without Doubt
If You Can't Change the Country, Leave the Country
Live Fearless with Fear
India: The Investment Magnet
Realignment
Powering a Billion Dreams
Universal Skill
Extension: From Brush to Algorithm

Watch for more at https://www.sandeepjchavan.com/.

www.ingramcontent.com/pod-product-compliance
Lightning Source LLC
LaVergne TN
LVHW052248200926
848481LV00002B/308